SYSTEMC

SystemC

Methodologies and Applications

edited by

Wolfgang Müller
Paderborn University, Germany

Wolfgang Rosenstiel
Tübingen University, Germany

and

Jürgen Ruf
Tübingen University, Germany

KLUWER ACADEMIC PUBLISHERS
BOSTON / DORDRECHT / LONDON

A C.I.P. Catalogue record for this book is available from the Library of Congress.

ISBN 978-1-4419-5361-2   e-ISBN 978-0-306-48735-4

Published by Kluwer Academic Publishers,
P.O. Box 17, 3300 AA Dordrecht, The Netherlands.

Sold and distributed in North, Central and South America
by Kluwer Academic Publishers,
101 Philip Drive, Norwell, MA 02061, U.S.A.

In all other countries, sold and distributed
by Kluwer Academic Publishers,
P.O. Box 322, 3300 AH Dordrecht, The Netherlands.

Printed on acid-free paper

Contents

Foreword

SystemC started with a vision of being more than a language definition. It also embraced the idea of a community of system level designers and modellers, sharing their ideas and models, and contributing to the evolution of both the language and the methodologies of how to use it. While the Open System C Initiative (OSCI) organisation is able to drive the language definition forward, it is the user community that is making strong contributions to SystemC modelling approaches and methodologies.

When we wrote our book on SystemC[1] in 2001–2, this user community was clearly emerging, with a number of very interesting presentations being made by a wide variety of industrial and academic users in major design and design automation conferences, and user group meetings. Most notable among these was the very strong European presence, especially as represented at the very active European Users group organised by Professor Wolfgang Rosenstiel of the University of Tübingen. This group has been holding twice–yearly meetings since early 2000.

This book, *SystemC – Methodologies and Applications*, represents the state of the art in system level design methods and modelling approaches for using SystemC in answering key design questions. The editors of the book, Wolfgang Müller, Wolfgang Rosenstiel and Jürgen Ruf, have done an excellent job in pulling together contributions from leading European groups into the volume.

When we consider what is needed to turn a language such as SystemC into an everyday part of system modelling and design practice, we come up with several requirements:

- Design methodologies for complex systems, involving combinations of hardware and software

- Modelling abstractions to allow creation of more effective system level verification

---

[1] Thorsten Grötker, Stan Liao, Grant Martin and Stuart Swan, System Design with SystemC, Kluwer Academic Publishers, 2002.

- Rigorous language semantics to allow the creation of appropriate design tools and to clarify modelling approaches.
- System level performance analysis approaches.
- Methods for improving designer productivity.
- Extensions to enable heterogeneous system modelling

The appearance of this book is rather timely from an industry perspective. SystemVerilog is emerging as a major standard for higher abstraction in HDL design, modelling and verification. It unifies design and verification capabilities (testbench, assertions, coverage etc.) in a single language and provides an implementation path from higher abstraction to gates. This offers a unique opportunity to build higher abstraction links between SystemC and SystemVerilog and unify transaction level modeling paradigms. Many key concepts of SystemC discussed in this book will facilitate such opportunities.

The eleven chapters in the book nicely match the above requirements. The first three chapters deal with applied design methodologies accompanied by case studies. The first chapter presents a comprehensive methodology for SoC modelling and design. Although complete automation was not possible, there are several advantages gained through the use of this structured design approach. The second chapter addresses the concept of "transaction level modelling" (TLM) and describes a SystemC based TLM modelling and analysis environment, demonstrated on several examples including a multimedia and ARM based platform. Important simulation speedup results are presented through using SystemC based TLM abstractions. Chapter 3 describes a refinement flow from high level SystemC towards implementable ANSI C targeted to an ARC based system. This refinement includes estimation strategies for SystemC based models to allow more optimal SW implementation code. The example used is an OFDM Demodulator.

Chapter four is a change of pace. Here, SystemC simulation semantics are formally defined using distributed abstract state machines. Such a rigorous definition will aid in creating advanced application tools, and determining and defining model interoperability with other languages. Two chapters exploring other aspects of system level modelling with SystemC follow this. Chapter 5 explores the issues of design error modelling and addresses emulation techniques interfacing SystemC to hardware emulators. Chapter 6 examines classical system level performance estimation, with SoC multi–processing architectures for networking applications. Their case study of a TCP/IP packet processing application and multi–processor platform indicate a significant speedup of simulation throughput with very low loss of accuracy, thus supporting higher level design space exploration.

The next three chapters deal with research aspects in the area of system level synthesis – automated transformations between different abstraction levels. A protocol specification language implemented as a proposed extension library to SystemC is discussed in chapter 7. The authors furthermore elaborate on an automatic translation method between communication abstractions as a means to develop automatic synthesis of protocol controllers. Chapter 8 addresses object–oriented hardware design and synthesis techniques while the following chapter deals with embedded software generation. It targets relatively simple hardware platforms using the eCos operating system as an example, and points to possible improvements in SystemC 3.0 to better support the SW modelling and code generation.

The final two chapters in the book deal with heterogeneous systems modelling, in particular, research carried out in the area of analogue and mixed signal extensions to SystemC. Chapter 11 concentrates on methods for design, simulation and refinement of complex heterogeneous signal processing systems using SystemC, with examples for extensions of SystemC for analogue modelling. It describes in detail the coupling between different design domains, and the refinement of a simple signal processing application.

The scope, depth and range of the work represented in these chapters indicate the vigour and vitality of the European SystemC community and are an excellent harbinger of more methodologies to come in the future.

This book is an extremely valuable addition to the literature on SystemC. Designers, modellers, and system verifiers will all benefit from the lessons taught by all the contributors. In the area of SoC design, SW development is clearly in the critical path. This is where SystemC can play to its strengths enabling designers to rapidly create system level abstractions of algorithms and architectures that have a vital role in system level design and analysis. They also provide the SW designer with an executable model of the SoC platform in order to start SW development and integration as early as possible. The fact that SystemC is based on C++ also helps to pave the road to SW.

This book will also be of great benefit to students and researchers who wish to understand more about how SystemC has been used, how it can be used, the underlying language semantics, and possible future evolution, in step with SystemVerilog. We hope this will just be the first in a long series of books reporting on user experiences and methodologies using SystemC, for the benefit of the entire SystemC community.

Grant Martin
Berkeley
Thorsten Grötker
Aachen

March 2003

Preface

We put great effort into the selection of authors and articles to present a high quality survey on the state of the art in the area of system design with SystemC. Organised into 11 self-contained readings, we selected leading SystemC experts to present their work in the domains of modelling, analysis, and synthesis. The different approaches give a comprehensive overview of SystemC methodologies and applications for HW/SW designs including mixed signal designs. We know that any collection lacks completeness. This collection mainly results from presentations at European SystemC User Group meetings (www-ti.informatik.uni-tuebingen.de/~systemc). We believe that it gives a representative overview of current work in academia and industries and serves as a state–of–the–art reference for SystemC methodologies and application.

Any book could never be written without the help and valuable contributions of many people. First of all we would like to thank Mark de Jongh, Cindy Zitter, and Deborah Doherty from Kluwer who helped us through the process. Many thanks also go to the contributing authors and their great cooperation through the last weeks. For the review of the individual articles and valuable comments, we acknowledge the work of Axel Braun (Tübingen University), Rolf Drechsler (Bremen University), Görschwin Fey (Bremen University), Uwe Glässer (Simon Fraser University), Daniel Große (Bremen University), Prakash Mohan Peranandam (Tübingen University), Achim Rettberg (C–LAB), Axel Siebenborn (Tübingen University), Alain Vachoux (EPFL), as well as many other colleagues from C-LAB, Paderborn University, and Tübingen University.

Wolfgang Müller
Paderborn
Wolfgang Rosenstiel, Jürgen Ruf
Tübingen

March 2003

* * *

Wolfgang Müller dedicates this book to Barbara, Maximillian, Philipp, and Tabea.
Wolfgang Rosenstiel dedicates this book to his family and the SystemC community.
Jürgen Ruf dedicates this book to his wife Esther and his children Nellie and Tim.

Chapter 1

A SystemC Based System On Chip Modelling and Design Methodology

Yves Vanderperren[1], Marc Pauwels[1], Wim Dehaene[2], Ates Berna[3],
Fatma Özdemir[3]

[1] *STMicroelectronics Belgium (previously with Alcatel Microelectronics)*

[2] *Katholieke Universiteit Leuven, Department Elektrotechniek–ESAT–MICAS*

[3] *STMicroelectronics Turkey (previously with Alcatel Microelectronics)*

Abstract   This paper describes aspects of the process and methodologies used in the development of a complex System On Chip. SystemC played a key role in supporting the technical work based on a defined refinement process from early architectural modelling to detailed cycle accurate modelling elements which enabled early co-simulation and validation work. In addition to SystemC, significant use was made of the Unified Modelling Language, and process and methodology associated with it, to provide visual, structured models and documentation of the architecture and design as it developed.

1.1   Introduction

The challenges presented to the design community by the ever greater potential of System On Chip technology are well documented [de Man, 2002, Scanlon, 2002]. SystemC provides the designer with an executable language for specifying and validating designs at multiple levels of abstraction. We decided to adopt the use of SystemC as an integral part of the design process for a recent System On Chip (SoC) development [1]. The product under development was a Wireless LAN chipset. In order to address a global market it was required to support more than one networking standard sharing very similar physical layer

---

[1]within the Wireless Business Unit of Alcatel Microelectronics, acquired by STMicroelectronics in 2002.

W. Müller et al. (eds.),
SystemC: Methodologies and Applications, 1–27.

requirements [ETSI, 2000b, IEEE, 1999b]. To ease integration of the devices into larger systems, significant higher level protocol complexity was required—in addition to the already complex transmit and receive signal processing in the physical layer. To efficiently and flexibly address the implementation of the signal processing functions of the physical layer an architectural approach was adopted which made use of a number of custom designed, microprogrammable processing engines in combination with general purpose processing cores. The scale of the project, both in terms of complexity and the degree of architectural novelty, demanded from the project team to adopt new approaches for the further development of the design in order to manage the risks involved. In addition to SystemC significant use was made of the Unified Modelling Language [OMG, 2001], and process and methodology associated with it, to provide visual, structured models and documentation of the architecture and design as it developed.

1.2   An Overview of the Methodology

The methodology presented was a fusion of some of the best ideas from the digital hardware and software engineering worlds.

The engineering team had significant previous experience of systems development using UML tools and associated methods. Other members of the team had long experience of digital hardware design using VHDL and recognised the importance of adopting the modelling and abstraction techniques offered by SystemC in order to tackle ever more complex designs. Key central ideas from this combined experience formed the foundation of our approach:

- **Iterative Development** — based on the ideas originally promoted by Boehm's spiral model [Boehm, 1988] and central to many modern software development processes [Rational, 2001]. The project is structured around a number of analysis, design, implementation and test cycles each of which is explicitly planned to address and manage a key set of risks. The project team maintains the risk list as the project unfolds and knowledge and understanding of the issues develops. Each iteration is targeted to resolve or mitigate a number of these, which may be technical risks associated with the design, or managerial and procedural risks such as finalising a new process, becoming familiar with new tools or achieving efficient working relationships with remote design teams. Testing evolves during each iteration in parallel with the system.

- **Use Case Driven Architecture** — the functional requirements of the *system* (not just the hardware or software components) are analysed in terms of Jacobson Use Cases [Jacobson et al., 1992]. This approach has been found to be very effective in several important ways:

- It enables a sound and structured statement of complex system requirements to be articulated—resulting in improved understanding across the team—and requirements validation with the customer or market representative.

- It drives the selection and assessment of the product architecture.

- It provides a foundation for the estimation, planning and management of project iterations, as well as an an excellent basis for system verification planning.

- **Proactive Requirements Management** — in a fluid requirements environment it is vital to maintain an accurate record of the project requirements. Modern tools make the task relatively straightforward, although the required manpower must not be underestimated. The tracing of functional requirements to Use Cases provides a powerful complexity-management technique and ensures straightforward review and approval of system verification specifications.

- **Executable System Models and Model-Centric Development** — the iterative design philosophy demands the ability to approach the design of the system as a series of explorations and refinements, each of which delivers answers to key questions and concerns about the design or the requirements. The ability to model the system and its environment at different levels of abstraction is vital to support such an approach.

The process presented here grew from these central ideas and is explained in more detail in the following sections. It should be made clear that this methodology did not attempt to modify the existing processes for detailed design and implementation. Rather it provides a comprehensive approach to the management of the transition from customer requirements through architectural design to subsystem specification (the entry point for detailed design) which is vital to control technical risks in a complex, multi-site development project.

1.3 Requirements Capture and Use Case Analysis

1.3.1 The Vision Document

One of the first system engineering activities was the drafting and review of the Vision Document. Though the Rational Unified Process [Rational, 2001] advocates such a document in a software engineering context, its value is no less for a *system* development. The Vision Document is the highest level requirements document in the project and provides the following:

- A statement of the business opportunity and anticipated positioning of the product in the market place and competitive situation.

4

- Identification of all anticipated users and stakeholders[2].

- Identification of the success criteria and needs for each user and stake-holder (in other words what do they need to get out of the project in order for them to consider it a success).

- Identification of the key product features and Use Cases. Features are the high level capabilities of the system that are necessary to deliver benefits to the users.

- Lists design constraints and other system level non-functional require-ments (e.g., performance targets), documentation requirements etc.[3]

The process of agreeing the contents of the Vision Document can prove cathartic. It is not uncommon for several different views of the project re-quirements to co-exist within the project and early alignment around a clearly-expressed set of goals is an important step. The Vision document forces con-sideration and alignment on fundamental drivers for the development and key goals.

1.3.2   Requirements Capture

Detailed requirements capture takes place from a number of sources. In the case of the Wireless LAN project, the primary source documents were the published networking standards to which the product needed to comply. The requirement database was built up from these documents and structured accord-ing to the features and Use Cases identified in the Vision document. Additional project requirements could be identified by consideration of product features and user needs which the standards do not address (such as field maintenance requirements, manufacture and test requirements, etc.). The resulting database allowed a number of requirements views to be extracted, from a comprehensive listing of all project requirements structured by feature to a high level feature list useful for discussing iteration planning.

1.3.3   Use Case Analysis

The Use Case Driven approach to software engineering was formalised by Jacobson [Jacobson et al., 1992], and has much in common with earlier ideas of system event-response analysis [Ward and Mellor, 1985]. Use Cases would normally be documented in textual form, augmented by sequence diagrams

---

[2]A stakeholder has an interest in the success of the project, but may not necessarily use the system directly.
[3]Key measures of effectiveness can be quantified here—non-functional parameters which are essential for product success (power consumption, battery life, throughput, cost) can be identified to focus attention.

showing the interactions between the system and its environment. UML provides a Use Case Diagram notation which assists in visualising the relationships between Use Cases and the system actors. Use Cases not only describe the *primary system response* (the expected behaviour of the system in order to satisfy the user goals), but also encourage early consideration of the *secondary system response* (i.e., what can go wrong in the successful scenario). It is important at this stage to avoid any assumptions about the internal design of the system—it remains the archetypal black box.

Use Cases will not hold all the requirements but only describe the behavioural part. This approach to the analysis of functional requirements is in no way restricted to software dominated systems, but provides a very general method for structuring functional requirements of any system. The requirements for a complete System On Chip design may therefore be analysed in these terms.

1.4   Modelling Strategies

As described in Section 1.2, modelling in various forms enables an iterative approach to systems specification. The primary deliverables of the modelling activity are proven subsystem specification and test environments as input to detailed design. Effort invested in creating the models is repaid by faster subsystem design and integration cycle times because the specifications contain less uncertainties and errors and are more clearly communicated.

1.4.1   Algorithmic Modelling

The requirements for the physical layer of the Wireless LAN system were heavily dominated by the need to specify signal processing algorithms. Although some of these are well known (e.g., Viterbi decoder), many required significant research and analysis. For such work an optimised toolset is necessary, and the team made full use of Matlab [MathWorks, 2000b] to develop and verify the algorithms. This work followed conventional lines and is not described further here.

Once verified, the Matlab model became the ultimate reference for the performance of the key signal processing algorithms.

1.4.2   SystemC Modelling

An Overview of the Physical Layer Architecture.   Although the focus of this paper is the methodology and process followed by the project, an overview of a limited part of the product architecture will ease some of the following explanation. Early in the development the following key architectural principles were developed:

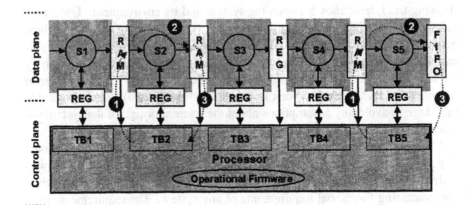

Figure 1.1. Physical layer architectural overview

- Certain complex signal processing functions should be implemented using custom processing engines designed specifically for the task. This approach provided a good compromise between the efficiency of dedicated logic and the flexibility of software based implementation for functions where the precise details of the algorithm could evolve and require changes.

- Signal processing stages would be isolated from each other via short FIFO queues, memories or register banks. This allows each section to run asynchronously with respect to each other and allows for very effective test and verification strategies since each section can be autonomously exercised from the control processor.

- Strict separation of the *user* or *data plane* and *control plane* handling should be observed. The user plane is implemented in dedicated logic or dedicated processors as described above to maximise the potential throughput of the design. Control plane functions are largely implemented in (hardware dependent) software on general purpose processor cores for flexibility and ease of development.

Figure 1.1 shows an overview of the physical layer architecture which emerged from these principles. The user plane information flows from left to right across the top of the diagram. User plane processing is carried out by the functions shown in ellipses. Each processing stage communicates with the control processor shown at the bottom of the figure. The control processor also has access to the data in each of the memory stages. The boxes labelled TB1 – TB5 and the dashed arrows show how each section can be independently exercised by dedicated test bench firmware using this architecture (refer to Section 1.5.3).

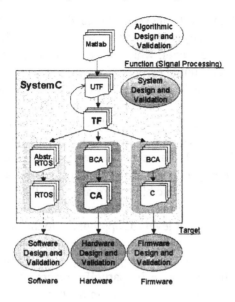

Figure 1.2. SystemC design flow

Model Development. SystemC was used to develop an executable model of the whole system architecture to gain confidence in the decisions which had been made. In the case of the physical layer, examples of concerns were:

- Would key sections be able to process data fast enough operating at critical clock rates?

- Would the chosen architecture implement the designed algorithms with sufficient accuracy to reproduce the performance obtained in the Matlab environment?

- The need to specify control and configuration interfaces in detail to enable control plane software to be designed.

overview of the abstraction levels: Figure 1.2 shows the general design flow a SystemC model adapted to take into account the peculiarities of the physical layer modelling (Matlab algorithmic work), and to make a distinction between high level software (running on RTOS) and hardware dependent software (also called *firmware*). The initial behavioural requirements are provided by Use Cases and, especially for signal processing oriented subsystems, by mathematical models in Matlab. The first modelling phase (Untimed Functional, UTF) represents each subsystem block at the functional level. Using this approach a model of the system can be constructed using a fairly high level of abstraction— with the benefits of speed of construction and execution—to gain confidence and insight on the high level subsystem interactions. A timed functional description of the system (TF) is obtained by adding information on throughput and latency targets to the model, which allows a broad timing overview to be constructed and to obtain initial estimate for required queue sizes, pipeline delays, etc.. In the case of the Wireless LAN project, the UTF modelling stage of the physical layer was skipped since the Matlab modelling was considered to have provided adequate support to the architectural partitioning. These models can be used as the baseline specification for detailed design in their own right. The SystemC model can be further refined to provide cycle accurate (CA) simulation of communication busses or taken as far as RTL equivalent cycle accurate simulation of logic blocks. At any stage the complete design can be simulated using a mixture of these different levels, providing flexibility to explore the design at different levels as appropriate to the implementation risk. Co-simulation of SystemC and VHDL is also possible with many tools. This may be carried out in order to verify project designed VHDL in the system environment, or in order to incorporate 3rd party IP delivered in the form of VHDL. Examples of co-simulations are presented in Section 1.5.3. Of course, as the level of abstraction is lowered the accuracy and confidence of the results goes up, but so does the time required for simulation. A SystemC model provides very useful flexibility to construct verification environments which effectively trade off these two concerns.

channel design, an example: The layered architecture of SystemC allows users to extend the language by defining their own communication channels. We improved the refinement process from timed functional to cycle accurate models by taking profit of such capability and by defining new primitive channels. One of these, called npram, can, for example, be used to model the memory elements between the processing sections of the architecture illustrated in Fig 1.1. This channel presents the following features:

- **support for multiple abstraction levels** — the npram channel can be used both at TF and CA levels. Multiple read and write accesses can occur during a simulation step at TF level. A single operation per simulation step is allowed at CA level for each port which is attached to the channel interface. The channel neither needs to be substituted nor requires a wrapper in order to refine the model.

- **generic container** — the type of information the npram can handle at TF level is extremely flexible. The user defines the exact capacity of each instance of the npram. There is no limitation about the kind of supported data types nor the number of elements. The instance shown on the left in figure 1.3 holds, e.g., an array A of 3 objects of a user defined class Symbol, an array B of 2 sc_int<8> and a bool variable C. This

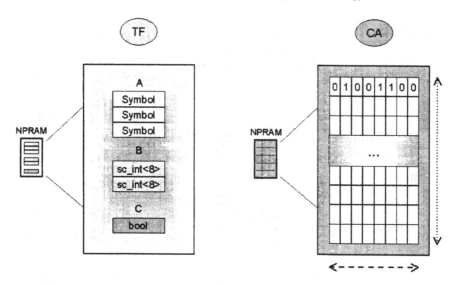

Figure 1.3. npram usage at different abstraction levels

flexibility allows the channel to be used in various contexts, without the limitation of indicating the type of data via template arguments, and without the constraint of a unique type per container instance. Moreover, by providing the description of the container structure outside the SystemC model, the user can, e.g., change at will the size of the FIFO queues and analyse the performances of the system in terms of quality of service under burst conditions. At CA level, the supported data type is restricted to bit true classes sc_int<> and sc_uint<>. The container is simply an array of bit true values, with a number of rows and word length specific to each instance.

- **configurable port registering** — the number of ports which can be connected to the channel interface, n, is instance-specific. Static design rule checking is performed to detect dangling ports.

- **verification of access conflicts** — the channel detects concurrent attempts to change the state of the container (i.e., writing at the same address at the same time). It is left up to the designer to specify how to resolve access contention.

It should be stressed that the npram channel is not limited to modelling memories. It provides a flexible container which may be implemented in various ways (e.g., register banks).

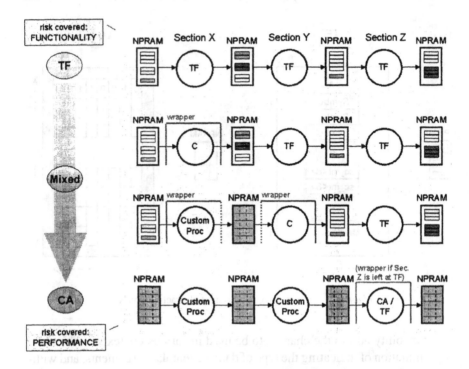

Figure 1.4. Sections modelling

sections modelling: As described in Section 1.4.2, the sections are to be implemented in dedicated logic, custom processors, or general purpose core, depending on the degree of risk associated with the function to be performed. In the former case the sections are modelled as much as possible at TF level, unless they are critical and require early and accurate assessment of their performances—direct CA level development. In the later cases the firmware can be encapsulated inside the SystemC model by means of proper layering. Depending on the complexity of the function to be implemented, the development of the firmware may follow an iterative development and be gradually refined, in a similar way to the management of the overall project (Section 1.5.1). While the custom processors were initially modelled explicitly and required low level code (assembler), solutions exist which allow to design application specific dedicated processors and cross-compile higher level code (C). Different situations may be simulated (fig. 1.4). In the first phase, only the behaviour of the *functionality* is modelled, with timing information, possibly using the complete range of SystemC instructions. The goal is to have as soon as possible an overall executable model of the system to verify the basic assumptions and key decisions about the architecture. This first step may be skipped in favour of early start of firmware design. The application code is developed, optimised, tested, and may be verified within the complete system. Compilation and linking occur on the host computer. In the next phase the Instruction Set Simulator (ISS) of the custom processor can be integrated. cycle accurate simulations of the cross-compiled application code are then performed. As said before and illustrated in figure 1.4, the complete system can be simulated at any stage using a mixture of the different levels, providing flexibility to explore the design at different levels as appropriate to the implementation risk. Encapsulation of firmware code within the SystemC model allows fast simulation but requires clean coding style, since the compiler of the host computer is used, not the target processor compiler. After cross-compilation the system *performance* can be evaluated with SystemC / ISS co-simulation. Depending on the level of refinement of the various sections, wrappers may be required for appropriate encapsulation. Indeed the `npram` does not perform conversion (e.g., in terms of data type or access details) between heterogeneous sections. As explained earlier, the same channel is used to model memories, register banks, FIFOs. The instances differ from each other at TF by their content (data types and sizes). At CA level the channels model arrays of bit true values.

fixed point design: In order to analyse the performance degradation of operating in floating or fixed point, a new data type called `fx_double` was used in the SystemC model which allows a single source code to be capable of simulating both floating and fixed point precision [Vanderperren et al., 2002]. Main benefits are:

- **fixed point design space exploration** — By providing the fixed point parameters outside the SystemC model, the fixed point design space exploration does not require successive time consuming recompilation steps.

- **inclusion of the scaling factor between floating and fixed point representation inside the model** — Figure 1.5 illustrates the scaling factor between floating and fixed point values. It can be used to verify the validity of arithmetic and relational operations, providing strong type-checking features—for instance, two fixed point numbers having different scales cannot be compared. The scaling factor is dynamically propagated along arithmetic operations—as an example, the scaling factor of the product of two `fx_double` numbers is the product of their scaling factors. Such feature is valuable, while modelling reconfigurable hardware (for instance a correlation with variable size), since each fixed point number is dynamically associated with the scale by which it must be multiplied in order to compare it with the floating point representation—the comparison between fixed point and floating point is straightforward.

- **absence of model conversion** — As a single model simulates both precisions, maintenance is less effort demanding. In the eventuality of func-

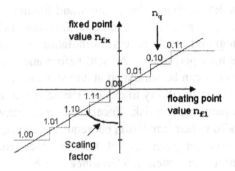

Figure 1.5. Scaling factor: a link between floating and fixed point values

tional bug fix, for example, there is no need to update both a floating and a separate fixed point model. This is also a major advantage in an iterative process where successive versions of the executable model with increased functionality are made.

- **reuse** — A single data type with dynamic specification of the fixed point parameters allows clean module interfaces and instances of the same module which can have different fixed point specifications.

Figures 1.6 and 1.7 show the interface of a given sc_module using the standard available data types and using the combined data type.

Of course these benefits come at the cost of simulation speed penalty of about 25% with respect to plain fixed point simulation. As most of this overhead is owed to the scale checking, this feature is switched off once the model is functionally tested and correct. The executable can then be used as a tool to explore alternative fixed point decisions with a speed penalty of 12.5%[4].

Though the fixed point design exploration was done manually in the context of the Wireless LAN project, this data type does not prevent to be used as a modelling support for automated fixed point exploration methods, as proposed in [Sung and Kum, 1995].

Summary of Experience. The following practical conclusions were drawn from the modelling activity:

- SystemC is just a language; a sound methodology and process is necessary to support its application in a project of significant size.

- Many disciplines familiar to software engineering must be applied to the development of the model, for example:

---

[4]Using g++ 2.95, SystemC 2.0, SunOS 5.7

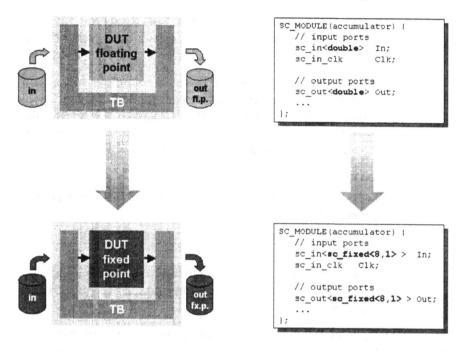

Figure 1.6. Fixed point design using the C++ and SystemC data type

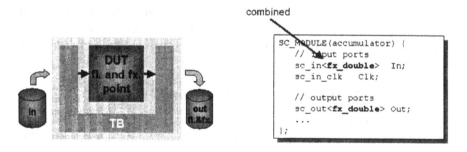

Figure 1.7. Fixed point design using the `fx_double` data type

- Clear coding standards and documentation style.

- Good control of overall architecture and interface definition between developers to avoid model integration problems.

- Careful attention to coaching and mentoring eases the adoption of new techniques (e.g., full use of the C++ features like the STL) and helps to build consistent approaches to design.

14

- An executable model results in fast feedback from designers, customers and marketing—allowing early bug detection and higher-quality design cycles.

- The effort spent developing generic solutions to recurring issues is paid back later by reuse and shorter design time. Collecting the results into class libraries enforces consistent use. This facilitates integration and the resulting uniformity assists maintenance and assimilation. These generic solutions share common principles with design patterns[5] and frameworks:

  - Interface is separated from implementation.
  - What is *common* and what is *variable* determines *interface* and *implementation*.
  - Variable *implementations* can be substituted via a common *interface*.
  - They embody successful *solutions* to recurring *problems* that arise when designing a *system* (not only software components) in a particular *context*.

It is not unimportant to note that the activity of building and testing the SystemC model serves as an excellent catalyst for cross-functional communication. The project deliberately involved design team members in the development of the model. This not only gave them deep familiarity with the content of the model—making the transition to detailed design much smoother—but also brought together representatives from different disciplines with the clear and tangible objective of modelling the system—encouraging cross-fertilisation of ideas and techniques.

1.4.3   UML Modelling of Architecture

UML was applied early in the project to the analysis of requirements and to the specification of higher layer software, using modelling and process guidelines based on the RUP [Rational, 2001]. The details of the software modelling approach are outside the scope of this paper.

As the SystemC model developed it became apparent that the process of developing and documenting the SystemC model would need to be carefully managed. The model was quite complex and needed to be accessible to a variety of people across the project. An architectural model was developed using a UML tool with the following objectives:

---

[5]However design patterns are often independent of programming languages and do not lead to direct code re-use. Frameworks define semi-complete applications that embody domain specific object structures and functionality: complete applications can be composed by inheriting from and/or instantiating framework components. In contrast, class libraries are less domain specific and provide a smaller scope of reuse.

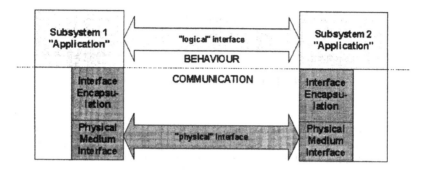

Figure 1.8. Model layering separating behaviour from communication

- To provide a common environment for the specification of high level architectural decisions across the disciplines.

- To specify the black box requirements for each major subsystem.

- To specify in detail the requirements for inter-subsystem interfaces.

- To provide a documented framework for the structure of the SystemC model—each major subsystem mapped to a sc_module in the SystemC code.

A modelling exercise of this scope required a robust structure to manage its development and to allow useful abstractions to be maintained. The following subsections discuss the most important principles established.

Use of UML Stereotypes. The UML provides mechanisms to extend and apply the language in specialised domains, allowing a UML model element to be distinguished as having special usage. Stereotypes to represent important concepts of structure (subsystem, channel, interface) were defined and supported the layering of the model described in the following section. Stereotypes representing SystemC concepts (<<sc_module>>, <<sc_port>>) allowed an explicit mapping between the UML and SystemC model.

Model Layering. An important abstraction adopted in the architectural modelling was the separation of high level domain-relevant behavioural modelling from the details of subsystem interfacing. Figure 1.8 illustrates this layering concept.

At the higher level subsystems interact with each other using 'logical' interfaces provided by the communication channels which connect them. This interaction occurs at a high level of abstraction and focuses on how the sub-

Figure 1.9. Logical view of subsystems and channels

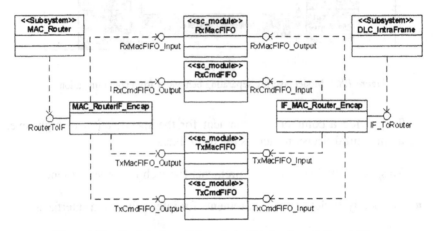

Figure 1.10. Design view of channel showing implementation entities

systems work together to meet the requirements—for example a typical logical primitive might be SendPDU(NetworkAddress, PduData).

The lower level of the model deals with the detailed design of the communications interfaces—whether the physical medium for communication should be a synchronous or asynchronous bus, shared memory or a point to point serial connection and what further processing is required in order to support the high level 'logical' interface over the chosen medium. A typical design level interface primitive might be WriteByte(RamAddress, ByteData). Figure 1.9 shows a typical structural diagram of the logical view of part of the system and Figure 1.10 shows a design level view of the same channel.

This technique allowed clean separation of concerns and detailed issues of software / hardware interaction to be considered within a framework which clearly captured the overall intent of the interface. The explicit identification in the model of hardware entities (for example, a dual port RAM) allowed associated design information (memory maps) to be directly linked to the model browser.

The model is developed by documenting Use Case Realisations using UML sequence diagrams. These realisations illustrate how the architectural elements interact in order to fulfil a given Use Case scenario. This is an extremely useful, although simple, method of illustrating dynamic interactions between elements of the model and allows details of low level interactions between software subsystems and hardware entities to be analysed.

Summary of Experience. The application of UML as a co-modelling tool provided many benefits. The most significant of these were:

- A common, structured environment for the documentation of system requirements and design information.

- Enhanced inter-disciplinary communication, like the consideration of sequence diagrams showing complex software-hardware interactions, reduces misunderstandings and disagreements between the disciplines earlier and more directly than would be expected with conventional specification approaches.

- Usefully separates abstraction levels to aid understanding.

Of course, there have to be some problems and issues, chief amongst these are:

- The architectural model is based on an a priori decision about the partitioning of functionality in the system. If the architecture is modified, the model needs to be manually brought into line (but then so would a textual document).

- There is no direct link between the UML model and the SystemC model. If the definition of a SystemC module is changed it is not automatically reflected in the UML model and vice versa. Many UML tools provide a round trip capability between the code and the model to ensure the two views remain synchronised and it would be very useful to explore the possibility of exploiting such capability for a SystemC model.

- Engineers without a UML or OOA background need to be introduced to the modelling concepts, process and notations. On the one hand this is a steep learning curve, on the other—if digital design engineers are going to move into the world of object oriented executable modelling (and it seems that they must)—a grounding in current best practices from the software industry can hardly be a bad thing.

A Step Further: the Error Control Example. The method and approach described in the preceding sections was developed during the project. Since its completion further work has been ongoing to explore possible improvements to the process for future projects.

One of the limitations of the approach described above was the need to select an architecture first and then begin the modelling process. If poor architectural decisions were taken at the beginning, significant rework of the models could be required. One of the strengths offered by SystemC is support for higher levels of abstraction in modelling. This section describes a short trial exercise which has been carried out to further assess the potential this capability offers to develop early untimed abstract models of the system. This allows the required functionality of the system to be modelled free from any assumptions or decisions about the implementation architecture. Once this model is constructed it can be used to explore and give insight to a number of architectural options.

The 'system' which was chosen as the basis for the trial, the Error Control [ETSI, 2000a], provided some challenges both in terms of understanding the complex requirements and in finding efficient architectural partitioning between hardware and software implementation.

The initial modelling phase has as its objective the construction of an executable model which provides the correct black box behaviour against the Error Control specification. This model should make no assumptions and place no constraints on the way in which the system will be implemented—known classically as an analysis or essential model [Rational, 2001, Ward and Mellor, 1985]. The reasons for constructing such a model are:

- It permits clear separation of concerns between the specification of what the system has to do and how it will be implemented.

- It allows systems requirements to be explored in detail and the understanding of the systems architects to be tested with the unforgiving sword of executable testing.

- It provides a mechanism for the formal specification of important algorithms to be developed and verified. For algorithms which are not particularly signal processing oriented (such as the buffer management algorithms for EC here) this environment is better suited to the task than Matlab.

- It allows the early development of a system level test suite, which can be reused as the design progresses.

Once again the use of a UML tool supported well the development of the SystemC model. Figure 1.11 shows part of the black box view of the system and its environment. Note that the UML icon capability has been used to enhance the readability of the diagram—SystemC ports are represented as squares, interfaces as circles and channels as boxes containing a double-headed arrow. The system interfaces are all modelled as SystemC channels, allowing the services provided by the channels to be separated from their implementation. The test bench module communicates with the system module via these channels.

The internal structure of the Error Control system (fig. 1.12) is implemented

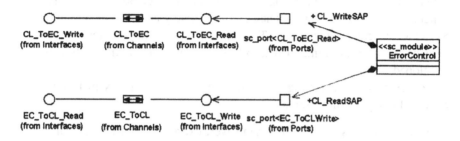

Figure 1.11. Part of Error Control UML model

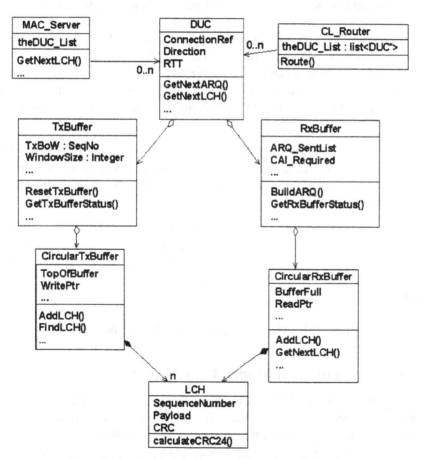

Figure 1.12. Partial class diagram of Error Control behaviour

as a set of C++ classes which provide the required functionality. These classes are developed using ANSI standard C++ and invoke each other's methods using synchronous procedure calls. They are deliberately coded to avoid any use of SystemC structuring concepts. This allows these 'analysis' level classes to later be mapped and remapped into SystemC modules as the architecture is developed.

The exercise of analysing the Error Control requirements in this fashion provided many useful insights:

- Complexities in the requirements were extracted much earlier than would otherwise have been the case. Domain and system experts can confidently specify important algorithms early in the project.

- The exercise of coding the required functionality allowed the consideration and discovery of clean and logical structuring of the requirements applying the classical principles of cohesion and coupling. A good example of this is the separation of concerns between the 'logical' problem of managing a notional 1024 element ring buffer, and the lower level problem of mapping this logical buffer to a physical area of memory with less than 1024 element capacity. The EC example code captures this distinction nicely and provides a useful basis for a possible architectural split.

- Very early consideration of system test scenarios is forced by the early exercising of the abstract model—not left to a late stage in the programme to be invented by a team who were never involved in the original conception and design.

The process outlined here is seen as a potential 'front end' to the material discussed previously. The use of an early analysis model allows a considered and informed approach to the adoption of an architecture. Once the architecture is selected, all of the techniques described earlier in the paper are, of course, available.

1.5 Iterative Development, Model Refinement and Verification

1.5.1 Iterative and Model Based Design

It is generally accepted in software engineering circles that one should strive to keep development iterations short. Too lengthy an iteration risks investing too much of the project resources, which makes it difficult to back track if better approaches become apparent in the light of the iteration experience. It is also important that the result of the iteration delivers output of real value in illuminating the way ahead for the team.

The Wireless LAN project iterations are shown in figure 1.13. The iteration plan may seem sluggish to software developers used to significantly more rapid cycles, but independent teams within the project were free to execute their own, more frequent iterations within the scope of a single project iteration.

The *feasibility* step covers risks related to algorithms using the Matlab platform and is aimed at assessing the signal processing performances. Iteration 0 concentrates on the mapping from the algorithms to an optimal *system architecture*. The obtained SystemC executable specification answers to key questions and allows starting the design process in the next iterations. In order to lower the risks associated with the transition from SystemC to VHDL, a limited VHDL activity already started during iteration 0. Once validated on a limited scope, the transition from SystemC to VHDL is applied to all the design. In a similar way a residual algorithmic activity may happen during iteration 0, covering remaining details which did not prevent starting off architectural brainstorming. During iteration 0 sections of the design which were considered high risk were modelled at CA level and co-simulated with TF level blocks. During iterations 1 and 2 (*detailed design*) a number of co-simulation scenarios were planned, involving the use of an Instruction Set Simulator to support co-verification of VHDL, SystemC, and target software. The SystemC model is used as a golden reference for the implementation and the *validation* of the final product during iteration 3.

1.5.2   Work Flows for Detailed Design

The modelling and simulation activity provides a verified, executable specification as input to detailed design of software, firmware, and hardware. The intent is that the model should be functionally correct. The design challenges to implement target software which is performant and meets the resource constraints of the target and detailed synthesisable RTL hardware definitions remain to be solved during the detailed design phase. The philosophy adopted during the project was that, at least at this stage of maturity of methods and tools, the translation from specification to detailed implementation must remain a manual design activity using tools and techniques appropriate to the discipline.

1.5.3   System Verification & Validation

Most early test activity deals with ensuring that the early architectural and analysis work correctly delivers the specified functionality and performance (verification). One of the strengths of the presented approach is the support it provides for the development of system verification environments (test benches and test scenarios) in tandem with the design of the system itself. These test benches are maintained and used throughout the project for system, subsystem and component testing (*vertical test bench reuse*). As illustrated in figure 1.14,

	S Feasibility, Requirements and Algorithms	It 0 Architecture and High Level Modelling	It 1 Detailed Design and Cosimulation	It 2 Hardware Prototype (FPGA)	It 3 Silicon
Systems	Capture reqts; Agree Vision Doc; Create UC model; Develop key algo's	Specify Architecture; Create HL SystemC / UML model; Demonstrate architecture meets project requirements	Cosimulation of SystemC, VHDL and SW	Cosimulation of SW on target and HW on FPGA; Conduct system V&V	Silicon Verification; Product Qualification or Approvals
Hardware		Involvement with SystemC specification	Detailed Design of VHDL	Port design to FPGA	Back end design and silicon fab
Software		Specify SW Arch; Create Host-Based SW framework and basic functionality	Build further SW functionality on It0 framework; Port to target	Build further functionality on It0/It1 framework	Build further SW functionality
Model or Platform used	Matlab	SystemC: Matlab as reference for floating point operations	VHDL & C: SystemC as reference for cosimulation	FPGA	Product on silicon
Test target	Function	Timing (TF mostly, CA for critical parts)	Check design vs. SystemC reference	Full system operation	Full real-time system operation

Figure 1.13. Wireless LAN iterations summary

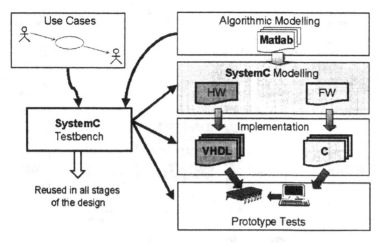

Figure 1.14. Vertical test bench reuse

the Matlab model serves as an algorithmic reference for the physical layer SystemC model, which in turn becomes a reference for the later stages of the design, on the path down to implementation. Test Benches may even be incorporated into product code to provide Built In Self Test functionality to the final product (*horizontal test bench reuse*). The control processor has a test mode, on top of its normal operation mode, which allows to test each section via 3 steps (figure 1.1): load test patterns (1); execute (2); obtain and analyse the results (3).

SystemC Verification from Matlab and Use Cases.   The Matlab model is the algorithmic reference for the physical layer system. In order to compare the performance of the SystemC model against the original algorithms, a bridge between the Matlab tool and the SystemC environment was required. The interface between Matlab and SystemC was achieved by using Matlab APIs [MathWorks, 2000a] from within SystemC. In this manner the input stimuli for the SystemC model were extracted from Matlab and output results of the SystemC model written to a Matlab file for comparison and analysis in the Matlab environment.

Figure 1.15 illustrates the framework for the SystemC verification. The bridge between Matlab and SystemC allows:

- A **seamless design development** from Matlab to SystemC, from algorithmic to architectural level.

- A **mathematical** and **graphical analysis environment** for SystemC.

The configuration file of the SystemC model allows the user to specify at will:

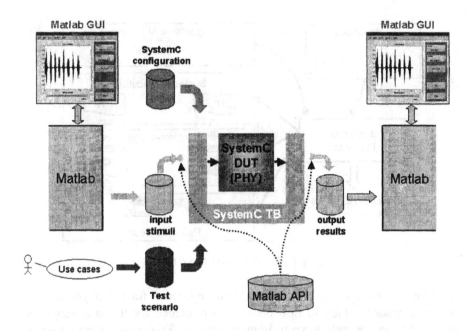

Figure 1.15. SystemC model environment

- **simulation parameters** — for example simulation duration, input and output file paths etc.

- **SystemC model parameters** — for example fixed point parameters, the size of the FIFO queues, the abstraction levels of the sections, etc.. As this information is kept outside the model, the *design space exploration* activity related to these parameters is decoupled from the *SystemC modelling* activity. This not only reduces time overhead (such as model recompilation) but allows flexible planning of distinct tasks and objectives.

The scenario file describes a given sequence of interactions between the system and its environment, corresponding to the realisation of a given Use Case Scenario.

The following paragraphs describe typical verification configurations which could be applied as the design progresses [Sayinta et al., 2003]. However the effort designing appropriate wrappers should not be underestimated. Obviously as the accuracy of the model increases, the simulation speed decreases.

SystemC with VHDL.   Figure 1.16 depicts the SystemC / VHDL co-simulation for the architecture given in figure 1.1. A section implemented as dedicated logic (S4) is replaced by its VHDL counterpart, in order to validate the implementation. The co-simulation tool supplies the environment to simulate the two languages.

SystemC with ISS. In figure 1.17 the SystemC model of the general purpose processor firmware is substituted with its Instruction Set Simulator (ISS) counterpart. This follows the same principles as explained earlier in the context of the custom processors. The aim is to verify the firmware implementation before any of the VHDL models is ready for the hardware part.

SystemC with VHDL and ISS. In the eventuality that the firmware is ready before the hardware parts are transferred into VHDL descriptions, a SystemC / VHDL / ISS co-simulation can be run (figure 1.18). Firmware code debugging can be started before all the hardware components are ready, in order to reduce overall design cycle time.

VHDL with ISS. This step validates accurately the hardware/firmware communication and follows the conventional lines.

System Validation. It is often the case with complex wireless systems that final proof of the system's capability of meeting customer expectations must wait until early prototypes can be tested in realistic environments (*system validation* activity). In the Wireless LAN project, the FPGA based prototype provides a pre-silicon platform which can be used to gain crucial confidence in the anticipated performance of the system and its algorithms in the real-world. It is important that controlled test environments (e.g. radio channel simulators) are considered for this important phase to allow the performance across the design envelope to be explored.

1.6 Conclusions

From its conception the project team recognised the need to adopt new and often untried practices in order to manage successfully the development of a complex design.

The combination of tools and techniques used for the development proved to support each other well and offer many advantages over traditional specification techniques. Executable modelling of the architecture allowed real confidence in the quality of specification to be gained early in the project—many specification errors were undoubtedly discovered and rectified much earlier in the process than would conventionally have been expected. On top of this the communication of the specification to the detailed design teams was smoother and less error-prone, both because of the formal nature of the specification and the involvement of the designers in the modelling process.

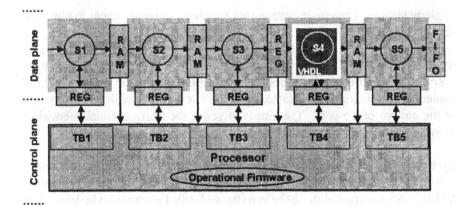

Figure 1.16. SystemC / VHDL co-simulation

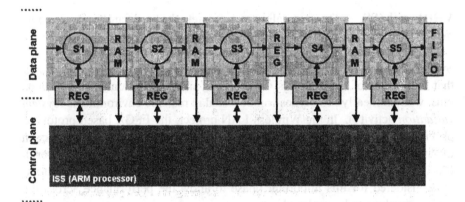

Figure 1.17. SystemC / ISS co-simulation

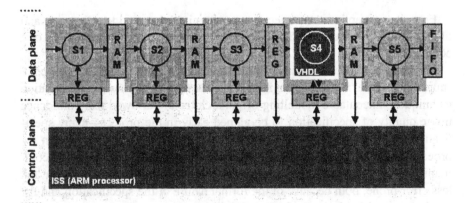

Figure 1.18. SystemC / VHDL / ISS co-simulation

Of course much work remains to be done to improve and streamline the presented foundations. Ultimately any development approach must satisfy a cost vs. benefit equation to the satisfaction of the organisation applying it. In the opinion of the authors the additional costs of the approach described in this paper are more than outweighed by the benefits of:

- **Improved product quality** — both in terms of meeting initial requirements and robust and flexible architectures capable of absorbing inevitable requirements changes.

- **Improved scheduling accuracy** — whilst it may be too early and over-optimistic at this stage to claim significant benefits in overall time to market for this approach, the team would certainly claim that the approach provides a much sounder basis for confident prediction of product release date than the traditional waterfall development.

- **Improved inter-disciplinary cooperation** — fostered by the activity of system level modelling and yielding real dividends in cross-fertilisation of ideas and better communications.

- **Improved confidence in the team** — the iterative approach provides many important psychological benefits to the team members. Early and regular demonstrations of progress in tangible ways not only gives management important confidence, but provides satisfaction to the engineering staff.

Acknowledgments

The authors would like to thank Jacques Wenin for his support during the project and his far sighted appreciation of the value of process improvement, and Trevor Moore for his valuable contribution to the presented work.

Furthermore, the methodology presented was derived from the work done in the Wireless LAN project; it would never have crystallized without the daily work and contributions of the system, hardware, and software teams in Belgium and Turkey.

Chapter 2

Using Transactional Level Models in a SoC Design Flow

Alain Clouard, Kshitiz Jain, Frank Ghenassia, Laurent Maillet-Contoz, Jean-Philippe Strassen
ST Microelectronics, France

Abstract   Embedded software accounts for more than half of the total development time of a system on a chip (SoC). The complexity of the hardware is becoming so high that the definition of the chip architecture and the verification of the implementation require new techniques. In this chapter we describe our proposed methodology for supporting these new challenges as an extension of the ASIC flow. Our main contribution is the identification and systematic usage in an industrial environment of an abstraction layer that describes SoC architecture to enable three critical activities: early software development, functional verification and architecture analysis. The models are also referred to as Transaction Level Models (TLM) because they rely on the concept of transactions to communicate. Examples of a multimedia platform and of an ARM subsystem highlight practical benefits of our approach.

2.1   Introduction

Multi-million gate circuits currently under design with the latest CMOS technologies not only include hardwired functionalities but also embedded software running most often on more than one processor. This complexity is driving the need for extensions of the traditional 'RTL to Layout' design and verification flow. In fact, these chips are complete systems: system on a chip.

Systems on chip, as the name implies, are complete systems composed of processors, busses, hardware accelerators, I/O peripherals, analog/RF devices, memories, and the embedded software. Less than a decade ago these components were assembled on boards; nowadays they can be embedded in a single circuit. This added complexity has two major consequences: (i) the mandatory reuse of many existing IPs to avoid redesigning the entire chip from scratch for

W. Müller et al. (eds.),
SystemC: Methodologies and Applications, 29–63.
© 2003 *Kluwer Academic Publishers.*

each new generation; and (ii) the use of embedded software to provide major parts of the expected functionality of the chip. These two evolutions lead to the concept of platform based design. As extensively described in the literature, a platform is based on a core subsystem which is common to all circuits that are built as derivatives of the platform. The platform is then customized for specific needs (specialized functionalities, cost, power, etc.) by either adding or removing hardware and software components. The embedded software accounts for more than half of the total expected functionality of the circuit and most often most of the modifications that occur during the design of a chip based on an existing platform are software updates. An obvious consequence of this statement is that the critical path for the development of such a circuit is the software, not the hardware. Enabling software development to start very early in the development cycle is therefore of paramount importance to reduce the time to market. At the same time it is worthwhile noticing that adding significant amount of functionality to an existing core platform may have a significant impact on the real time behavior of the circuit, and many applications that these chips are used in have strong real time constraints (e.g. automotive, multimedia, telecom). It is therefore equally important to be able to analyze the impact of adding new functionality to a platform with respect to the expected real time behavior. This latter activity relates to the performance analysis of the defined architecture. The functional verification of IPs that compose the system as well as their integration has also become crucial. The design flow must support an efficient verification process to reduce the development time and also to avoid silicon re-spins which could jeopardize the return on investment of the product under design. We will also describe how the proposed approach strengthens the verification process. At STMicroelectronics one direction to address these two issues is to extend the CAD solution proposed to product divisions, known as Unicad, beyond the RTL entry point; this extension is referred to as the System to RTL flow. As the current ASIC flow mainly relies on three implementation views of a design, namely the layout, gate and RTL levels, the extended flow adds two new views: TLM and algorithmic. In the remainder of this chapter we first introduce the System to RTL design and verification flow with its objectives and benefits. We then focus on the TLM view with the description of the modeling style and also the usage of these models to support critical activities of platform based design: early embedded software development, early architecture analysis and reference model for the functional verification. Finally, we present how we used this approach for the design of a multimedia multiprocessor platform as well as the modeling approach of an ARM based subsystem.

2.2 Overview of the System to RTL Design Flow

Our system to RTL flow relies on three abstraction levels (see Figure 2.1):

- *SoC functionality* models the expected behavior of the circuit, without taking into account how it is implemented;

- *SoC architecture* (i.e. SoC-A or SoC TLM platform) captures all information required to program the embedded software of the circuit;

- *SoC microarchitecture* (i.e. SoC-MA or SoC RTL platform) captures all information that enables cycle accurate simulations. Most often, it is modeled at the register transfer level (RTL) using VHDL or Verilog language. Such models are almost always available because they are used as input for logic synthesis.

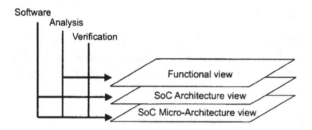

Figure 2.1. Overview of the SoC design flow

All views have complementary objectives and indeed avoid the usual conflict between the need for accurate descriptions that lead to slow simulation speeds and the request to simulate real time embedded application software. The SoC microarchitecture view is aimed at:

- Low level embedded software debug and validation, in the context of the real (simulated) hardware. Our goal is to enable the debug of the device drivers and their integration into the target operating system before the first breadboard or even the hardware emulator is available;

- SoC functional verification. We aim at verifying that the IPs, once integrated, still provide the expected functionality and that the parallel execution and communication of the blocks do not corrupt their behavior. Some blocks are processors which run software. The verification activity must take this fact into account and validate all possible interactions between the hardware and software components;

- SoC microarchitecture validation. In order to sustain the real time requirements of the application, the embedded software is usually optimized and the hardware configured accordingly. If performance is not sufficient, then the SoC architecture may be upgraded to match the requirements. The part of this activity that requires cycle accuracy relies on the RTL view.

An illustration of the above activities can be found in [Chevallaz et al., 2002]. The SoC architecture view is aimed at:

- Early firmware development. It enables the hardware dependent software to be validated before it runs on a prototype;

- SoC architecture exploration. In the early phases of the design cycle, architects have the task to define the architecture that will best suit the customer requirements. The early availability of the TLM platform is a means to obtain quantitative figures that are preceeded to make appropriate architectural decisions;

- Functional verification. Providing a reference model for the verification engineer. This 'golden' model is used to generate tests which are applied on the implementation model to verify that the functionality matches the expected behavior.

Finally, the SoC functionality, as its name implies, specifies the expected behavior of the circuit as seen by the user. It is used as an executable specification of the functionality, and is most often composed of algorithmic software. Performance figures are specified separately, often as a paper specification.

The proposed views can be integrated gracefully into the SoC development process, as explained below. The functionality is independent of the architecture so that its corresponding view can be started as soon as there is a need for a product specification.

While software and architecture teams are using the same reference SoC TLM model for firmware development and coarse grain architecture analysis, the RTL development takes place, leading to a SoC RTL Platform, as depicted in Figure 2.2. By the time this platform is available, some tasks more closely related to the hardware implementation are ready to start: fine grain architecture analysis, hardware verification and low level software integration with hardware. These tasks are conducted concurrently with the emulation setup, the synthesis and back end implementation of the standard ASIC design flow.

In the end, when the first breadboards are ready, the software (drivers, firmware and a simplified application) is also completed with good confidence and the SoC has gone through a thorough verification process that increase the probability to achieve first time silicon/system success.

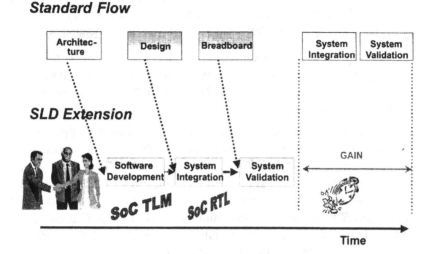

Figure 2.2. Concurrent HW/SW engineering

A current difficulty of the proposed approach is that in the absence of synthesis tools that would generate these views out of a reference model, the coherence with the two additional views needs to be maintained during the development of the circuit. We address this issue by reusing the functional subset of 'system' test vectors across all views, therefore ensuring their conformance to the expected functionality. Another possibility is to run the same embedded software (with no modification) on the different simulation environments. Identical behavior proves the compatibility of the underlying models. Obviously the coherence is ensured to the level of functional coverage that the tests provide. Also, as mentioned earlier, TLM models may be used as executable specification of SoC for RTL verification.

2.3 TLM, a Complementary View for the Design Flow

2.3.1 TLM Modeling Concepts

When addressing the need of early embedded software development and architecture analysis, engineers list the following requirements as enablers for their activity:

- *Speed*: models used for the above activities require millions of simulated cycles in some cases and it is not acceptable to wait for even a day to complete a simulation run, especially in the case of interactive sessions!

- *Accuracy*: some analysis requires full cycle accuracy of the models to obtain reliable results that can be trusted. A minimum requirement is that the model should be detailed enough to run the embedded software;

- *Lightweight modeling*: if required, any modeling activity besides the RTL coding (which is mandatory, anyway, for silicon) must be light enough so that its cost is not perceived as higher than the resulting benefits by SoC project management.

Complex designs usually start with the development of an algorithmic model (for example, based on a data flow simulation engine for DSP oriented designs). Of course, its advantage is that it simulates extremely fast because it only captures the algorithm, regardless of the implementation detail. However, such a model has no notion of hardware and software components, and does not model registers nor system synchronizations related to the SoC architecture. It therefore defeats the purpose of enabling the execution of the embedded software and is consequently not suited to our needs.

With the emergence of C based dialects that supported hardware concepts, it is often believed that writing cycle accurate (CA) models with a C based environment could respect all of the above requirements. This approach [Haverinen et al., 2002, Gerlach and Rosenstiel, 2000, Semeria and Ghosh, 1999, Grötker, 2002, Fin et al., 2001a] has several major drawbacks:

- Much of the information captured in such a model is not available in the IP documentation but only in the designer's mind ... and in the RTL source code. It resulted in much time either spent by the RTL designer updating the engineer writing the corresponding cycle accurate model or forcing the modeling engineer to reverse engineer the RTL code. Either way, it ended up being a tedious and heavy process, with equivalence issues always present;

- The cycle accurate model ended up being only an order of magnitude faster than the equivalent RTL simulation very similar to the speed of cycle based VHDL/Verilog.

Not only was the speed too low to run a significant amount of embedded software in a reasonable time but also the development cost became too high compared to the benefits.

At the same time we also noticed that architects and software engineers do not require cycle accuracy for all of their activities. For example, software development may not require cycle-accuracy until engineers work on its optimization. On other occasions they do not need high simulation speed together with cycle accuracy, e.g. when verifying cycle dependent behavior.

Based on this experience and new understanding, we adopted the famous 'divide and conquer' approach by providing two complementary environments:

- SoC TLM for early usage with a relatively lightweight development cost. This abstraction level sits in between the cycle accurate, bit true model and the untimed algorithmic models and is, in our opinion, the adequate trade off between speed (at least 1000 times faster than RTL in our experience, as depicted in Figure 2.3) and accuracy, when complemented with a SoC RTL platform;

- SoC RTL platform for fine grain, cycle accurate simulations at the cost of slower simulation speed and later availability.

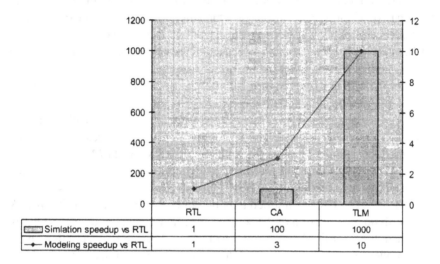

	RTL	CA	TLM
▨ Simlation speedup vs RTL	1	100	1000
──◆── Modeling speedup vs RTL	1	3	10

Figure 2.3. Efficiency of modeling strategies

Let us now define the terms and notions required to understand the TLM modeling approach. A system provides functionality with adequate performance to a user. It is composed of a mix of hardware and software.

A SoC is a special case of a system in which both hardware and software are embedded into a single circuit. A system model is composed of a set of elements, named modules, which:

- Execute part of the expected system behavior thanks to one or several concurrent threads;

- Communicate data between each other (and potentially with the test bench) to perform the system functionality. This may be a character, data of a full image, etc.. Each data set exchange is called a transaction. Two consecutive transactions may transfer data sets of variable size. The size corresponds to the amount of data being exchanged between two system synchronizations, as defined below;

- Synchronize between each other (and potentially with the test bench). System synchronization is an explicit action between at least two modules that need to coordinate some behavior distributed over the modules that rely on this synchronization. Such coordination is required to ensure predictable system behavior. An example of such synchronization is an interrupt raised by a DMA to notify a transfer completion.

The simulation speed up achieved by a TLM model compared to the equivalent RTL model has a direct relation to the ratio of the mean number of cycles between two system synchronizations which therefore includes one transaction. The presence of explicit system synchronizations is a mandatory condition for a TLM model to behave correctly because it is the only means of synchronizing all the concurrent events that occur in the system. In our experience, all systems comply with this rule. Figure 2.3 provides approximate ratios in terms of simulation speed up and modeling efforts.

For specific purposes it might be needed to refine a TLM model to include some information that relates to the microarchitecture. In this case we name the initial model that only captures the architecture TLM-A (TLM Architecture); the refined model is named TLM-MA (TLM Microarchitecture). Of course, for efficiency reasons a methodology is provided to ensure that the source code for the TLM IP can be either compiled to build either the TLM-A or the TLM-MA model with no modification of the source code. In this chapter, unless explicitly mentioned, the term TLM corresponds to TLM-A. In terms of abstraction levels such a model lies in between the architecture and microarchitecture levels, and its precise definition is therefore much more fuzzy.

2.3.2   Embedded Software Development

Software development (or at least debug and validation) can only start when it can be executed on the target platform. The traditional approach considers the physical prototype (emulator or FPGA board prototype) as the starting point for software development. The obvious advantage is that it becomes available much earlier than the chip. The drawback is that all the hardware development is nearly completed. Any need to modify the hardware uncovered by the execution of the software is consequently very costly. Hardware/software co-verification can be used to start executing software on the target (simulated) hardware even earlier. With this approach, any light hardware modification is still possible if this technique is used with the first top-level VHDL/Verilog netlists. The drawback of this approach is that it relies on the VHDL/Verilog models for simulation, resulting in slow execution speed that only enables reduced parts of the embedded software to be run. Also, major hardware modifications are too costly because the design is too advanced.

SoC TLM platforms can be delivered shortly after the architecture is specified. An important consequence is that it becomes available for software development very early in the hardware design phase. True hardware/software co-development is then possible. Hardware and software teams can work together using the TLM platform as the 'contract' between the two teams. The TLM models then become the reference models that software developers use to run their code. The same models are used as golden reference models by the hardware teams.

2.3.3   Functional Verification

Block-level Verification.   Functional verification aims at ensuring that the implementation of an IP complies with the specified functionality. Based on the paper specification, the verification engineer defines test scenarios to apply to the VHDL IP model. The engineer often also 'manually' determines the expected result of each scenario.

As discussed in the previous paragraph, a TLM model is the actual functional specification of the IP. It can also be referred to as the executable specification of the IP because it only captures the intended behavior, as seen from the 'user' of the IP but does not include any detail on the internal implementation. In other words, it models the architecture of the IP, not its microarchitecture.

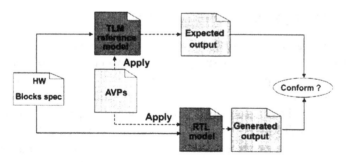

Figure 2.4.   Functional verification with TLM models

The TLM model can therefore be used in place of the manual process undertaken by the verification engineer to generate expected results, as seen in Figure 2.4. Applying the test scenarios on the TLM model will automatically generate the expected (also named reference) output. Of course, a recurrent question is how to make sure that the executable specification complies with the written specification, and therefore what is the level of confidence that the generated output can be used as reference. There is no answer to this question. However, the same question holds for the reference output when manually generated by the engineer. What guarantees that he did not misunderstand the specification or made a mistake while writing the reference output? Also, beyond the automation of the reference output, another benefit is to have a written specification of what has been used to generate this reference data so that in case of mismatch between the reference data and the one generated by the design, it is always possible to compare both source code and understand which one corresponds to the intended behavior.

SoC System Level Verification. In a typical design flow, once IPs are verified independently, they must also be verified in the SoC context. Functional integration tests are developed to verify features such as:

- Memory map: to avoid registers to be mapped at similar addresses and also to ensure that hardware and software teams rely on the same mapping;

- Data consistency: the data generated by one IP matches the input (format, content) data of the IP that will reuse the generated data;

- Concurrency: the concurrent execution of all IPs must be controlled to ensure deterministic and expected behavior.

The same approach as for block level verification holds. The SoC TLM platform is used to generate expected system reference output for a given scenario.

2.3.4 TLM Models for Legacy IPs

The above paragraph describes how to use a TLM model to verify the RTL implementation. An implicit assumption is that whenever an IP is designed, the TLM model exists prior to the RTL model.

A strict methodology can certainly enforce this sequence for new IPs. However, legacy IPs will most probably not comply with this approach and therefore will not include a TLM view. Most often a VHDL test bench will be delivered with the RTL model. The test bench also includes test vectors and reference output.

In the industry all designs have a significant level of IP reuse. Integrating the TLM view in a design flow is therefore not only a matter of promoting a new clean approach but also of proposing an evolutionary path from existing design methods. In this context it is necessary to develop TLM models for existing IPs. In this case the reference model is the RTL. The scenarios and associated test vectors have to be extracted from the RTL test bench, recorded, and then applied on the new TLM model in order to ensure consistency between the two views. After the TLM view of the IP is validated against the RTL, it can be integrated into a SoC platform for software development and architecture analysis.

Two cases must be taken into account while recording the activity between the design under test and the test bench:

- Data exchanges over a bus: bus monitors must extract the transactions and associated attributes (see subsection 2.3.7);

- Standalone signals (i.e. interrupts) are merely recorded.

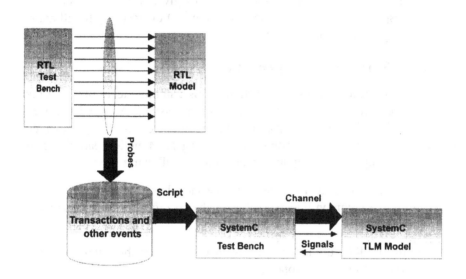

Figure 2.5. RTL as golden reference model

A SystemC test bench will then load the trace of transaction and signal values to drive the TLM model. The output is then compared with the reference data from the RTL test bench. Of course, this approach is limited to test benches that do not need to dynamically adapt the generated test vectors to some response from the design under test.

While reusing test vectors that were not initially built to run on a pure functional model, some limitations may force manual intervention or more sophisticated comparison methods to be implemented:

- Absolute timing: some IP behaviors rely on absolute timing information. TLM models do not capture cycle accurate timing information and will therefore likely not provide the same output sequence. A classical example is a timer;

- The test bench relies on some timing delays to trigger operations. In this case the TLM model will probably not execute the same number of operations during the time delay;

- The test bench is written in a way that assumes exact delays between events instead of relying on a partial order of events. Let us look at the example of a DMA that triggers an interrupt after it completes a data transfer. A 'clean' test bench will rely on a status register to identify the end of data transfer and then look for a value change of the interrupt. In this case the same test bench can be used as is on the TLM model. If the interrupt signal value is checked after a given number of cycles (that corresponds to the completion of the transfer) the test bench will not be directly reusable on the TLM model.

2.3.5 Early System Integration

System integration relates to the activity of assembling pieces of the system together to ensure the correctness of the resulting system. A stepwise process is usually adopted with successive partial integrations during the development of the hardware and software components. Using TLM models participates in the strengthening of the integration process in the following ways:

1 Software is executed on the reference models of the hardware IPs. It ensures a common understanding of the specification of the hardware blocks;

2 Hardware RTL implementation is checked against the reference model used by software developers;

3 Earlier integration between hardware components can take place by mixing TLM and RTL views of several IPs being integrated. This avoids the need to wait for all RTL blocks to be released to start any hardware integration activity;

4 Software can be executed on partial RTL platforms by using TLM models for IPs whose RTL code is not available yet. It enables earlier hardware/-software integration.

Mixed TLM/RTL co-simulation is an important feature of the design flow for supporting the above activities. Also, abstraction level adapters for interfacing TLM and RTL models are needed. For this purpose bus functional models can often be reused or adapted from a library of adapters provided with each communication protocol such as STBus or AMBA. Signals used for system synchronization are directly mapped from the TLM interface to the corresponding signals of the RTL IP model. Some RTL signals that are neither a system level

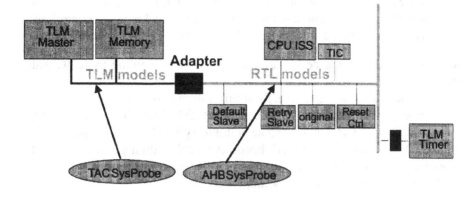

Figure 2.6. Mixed mode simulations

synchronization nor part of the bus protocol (e.g. reset, clock) have no direct equivalent in the TLM model. A specific stub (either in VHDL, Verilog, or SystemC) may be required.

2.3.6 Performance Analysis and Verification

Great care should be taken to ensure that the circuit respects its real time requirements. These constraints must be taken into account from the initial definition of the system. A verification process must take place at each step of the design flow to confirm that the constraints still hold.

The performance of the system can be studied as soon as the model includes timing information. Two cases may arise:

- Timed TLM-A model: it might be sufficient to add delays to execution paths of a TLM-A model to accurately mimic its timing behavior. The computation of the delay might vary from being straightforward with a fixed value to a complex function whose returned value varies over time and depends on information collected at run time. An illustration of this technique can be found in section 2.7.

- Timed TLM-MA model: simply annotating a TLM-A model might result in unacceptable inaccuracies. This often occurs for complex IPs. In this case, it is necessary first to refine the model into a TLM-MA model, as discussed in section 2.2. Timing information is then annotated. For example, it often necessary to refine an architectural IP model when it represents a shared resource. An arbitration policy must be added in place of the simple 'first in first out' policy[1] of the TLM-A model. Typical models which fall into this category are busses and memory controllers. The subsection 2.8.2 illustrates this approach.

In both cases the timing delays are modeled with calls to the SystemC wait statement.

Depending on the design phase, the timing information can either come from the architect or from analysis of the equivalent RTL simulation. For a true top down flow the architect, based on some preliminary benchmarking experiments, will initially insert it. Once the RTL becomes available, timing information will be extracted from the RTL simulation and used to annotate the equivalent TLM model. This activity is called back annotation.

2.3.7 Analysis and Debug Tools

The most critical tool is a standard debugger to control the simulation (run, step, continue, etc.) as well as display the values of variables of the executed program. Beyond these basic features it is important that the debugger is aware of the concurrent processes of the system so that the debugger enables a fine control on processes, signals and events but also allows developers to 'cycle step'. To fully understand the behavior of a parallel system (on chip), a trace viewer is also mandatory. As described below, such a viewer is not only powerful for functional debugging but also for performance analysis. A single analysis tool which supports all the SoC phases is needed. An example of such a tool is SysProbe[Clouard et al., 2002, Nikkei, 2001].

SysProbe is built around a commercial tool and basically:

- Records read/write transactions during RTL or TLM simulation into a database of transactions. If RTL, replaying a lengthy simulation is not necessary since SysProbe can detect transactions from a VCD file instead of during simulation. Of course, this flexibility sometimes has to be traded off with the drawback of extensive trace files being generated.

- Allows visualization of transactions over time and eases creation of statistics tables (see Figure 2.7). The upper lines are usual RTL signals. Rectangles are bus read/write transactions, with their parameters address, data value, duration, etc.. SysProbe extracts transactions from RTL simulations by monitoring signals of the bus control lines. For TLM simulations transactions are natively modeled, and therefore directly recorded. The table at the lower part of the graph displays computations from the database of recorded transactions. Alternatively to tables, SysProbe anal-

---

[1] Such policy is optimized for speed because is removes the need for a delta cycle

ysis results may be displayed in other diagram formats: pie charts, histograms, etc. (see Figure 2.7).

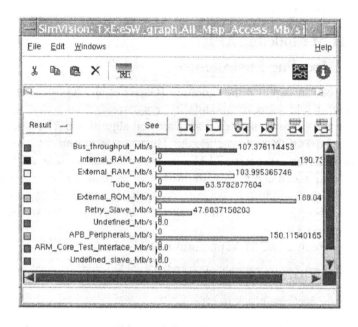

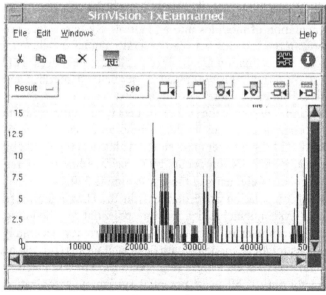

Figure 2.7. Examples of SysProbe analysis results

In addition, SysProbe enables the user to see which lines of embedded software source code have created specific transactions which the user has selected in the transaction viewer. This is a powerful 'hardware back to software' bug tracking and analysis feature.

2.4   TLM Modeling API

2.4.1   Modeling Environment

As described in depth in the literature [Gajski et al., 2000], our modeling environment should support: modularity, concurrency, hierarchy, separation of behavior from communication, hardware and algorithmic data types. After evaluation of C/POSIX and synchronous languages environments, which had some drawbacks for our specific TLM goal, we evaluated and selected SystemC [OSCI, 2002a], which complies with most of these requirements. In addition, becoming a *defacto* industry standard, SystemC is supported by most CAD tools, hence facilitating a smooth transition to successive steps of the design flow (see section 2.9).

Each module of a SoC TLM platform is characterized by:

- Bit true behavior of the block;

- Register accurate block interface;

- System synchronizations managed by the block.

For the integration of modules into a platform, SystemC does not offer the ability to perform transactions between several modules connected to a single channel based on a destination identifier (e.g. address map in case of bus based architectures). To address this we created a user defined interface and provided associated bus channels, as described in the next section.

The transaction size equals the size of the data type multiplied by the number of elements to transfer. For example, assume an image of size 640×320 described in RGB24 (24 bits per pixel stored with a data type of 32 bits integer) and that an image transmission occurs by successive line transfers. For each transfer the number of elements of the transaction is 320.

The model of computation which derives from the TLM definition (see above) implies that all system synchronizations are potential re-scheduling points in a simulation environment in order to model concurrency accurately. System synchronizations might be modeled with dedicated means (e.g. events, signals) but can also rely on data exchanges (e.g. polling). The simulated system will behave correctly only if all these potential system synchronizations cause a call to the simulation kernel, therefore enabling its scheduler to activate other modules. SystemC events and signals comply with this requirement. However, it is not possible for the simulator to distinguish transactions modeling data exchanges from implicit synchronizations. We must consequently adopt a conservative approach and allow the kernel to re-schedule after the completion of each transaction.

After the TLM platform is integrated execution of embedded software is possible either compiled natively on the simulation host, for highest simulation speed, or cross-compiled to the embedded processor architecture for execution on ISS (instruction set simulator), as described in the next section.

2.4.2   Modeling API

This section discusses the definition of a programming interface for IP models. We focus on defining an interface to support data exchange between IPs. Mechanisms to ensure IP synchronization (e.g. signals, events) are not discussed because they rely on standard mechanisms proposed in hardware design languages.

General Requirements.   SoCs are composed of many IPs. Building SoC simulation platform consequently requires the integration of many models. These IP models are probably developed by different design teams and may come at different abstraction levels (IP-A or IP-MA). Also, IP-MA models may rely on different communication protocols, as shown in Figure 2.8.

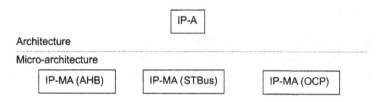

Figure 2.8.   Abstractions and protocols

General requirements:
The IP interface should enable the following activities:

- *Reuse of IP-A models:* architectural IP models should be plugged into a SoC-A model and ready to simulate with no code modification, regardless of their microarchitectural implementation. A consequence is the ability to combine IP-A models whose implementations do not use the same protocol (i.e. AHB, STBus, etc.);

- *Ports:* it should be possible to identify the ports used to send/receive data to/from the communication structure. Some IP behaviors rely on this information. For example, a memory controller may use this information to adapt its decoding scheme or priority of the incoming request. Having a field of the transaction structure indicating some information such as priority will not provide the same functionality. An example is the case of the embedded software controlling the priority level of any request going through a specific memory port;

- *Separation communication/behavior:* IP behavior should be completely separated from the communication mechanism, enabling various communication schemes to be used with the same IP-A model;

- *Information locality:* information should be captured in the model that 'owns' it;

- *Simulation efficiency:* usage of API should not result in 'significant' simulation degradation;

- *Intuitive API:* enable intuitive usage of API for required modeling activity so that learning curve is minimized and potential misuse avoided;

- *A2MA and MA2A adapters:* minimize the number of abstraction adapters to be developed;

- *Protocol adapters:* minimize the number of protocol adapters to be developed;

- *Automation:* it should be possible to automate activities of the use scenarios (see previous section) at minimum cost (i.e. tool development cost). Obvious automation is monitoring, adapter insertion, code generation and consistency checks.

IP-A vs. IP-MA:

Data types: the granularity of the data that is sent/received by IP-A models corresponds to the functionality of the block (e.g. image macro-block, ethernet packet). On the contrary, IP-MA models exchange data set that can be handled by the communication layer (e.g. 32 bits, burst of 32 bits). When an IP-A model exchanges data with an IP-MA model an adaptation layer describing how the data is 'chunked' into pieces (e.g. using unitary transfers, one or several bursts, packets, etc.). This adaptation is part of the abstraction adapter.

Control: IP-A and IP-MA models differ in the communication control layer they expect. An IP-A model only expects to read and write data whereas an IP-MA model expects to control the handshaking mechanism that is necessary for transfering the data. In this respect the control sequence is more complex for microarchitectural (MA) models.

Timing: no timing information is necessary for IP-A models. They only capture the sequence of events occurring in the system as the data computation. Apart from simple IPs (i.e. IP-A and IP-MA are almost identical), IP-A models are not suited to represent timing behavior.

On the contrary, IP-MA models capture how data is exchanged and also the internal structure of the IPs. They are consequently sufficiently detailed to enable accurate timing annotation.

TLM API.

TLM transport:
The TLM API enables the transport of transactions from IPs to IPs. A transaction is used to exchange information between 2 or more IPs. A transaction might involve bi-directional exchange of information.

A transaction conveys the following information:

- Initiator data set: data sent by the initiator of the transaction exchange;

- Target data set: data sent back by the target of the transaction exchange;

- Initiator access port;

- Target access port.

A corresponding structure follows:

```
class tlm_transaction {
  tlm_transaction_data *InitiatorData;//filled by initiator
  tlm_transaction_data *TargetData;   //filled by target
  sc_port_base *initiator_port;  // filled by
                                 // transaction mechanism,
                                 // available to target
  sc_port_base *destination_port;// filled by
                                 // transaction mechanism,
                                 // available to target
};
```

A transaction is initiated when an IP issues a request onto an access point. It completes when the IP resumes its execution. The requesting execution thread is suspended until the transaction completes. A function named transfer is called to initiate a transaction transfer. A C++ function prototype looks like:

```
tlm_status transfer(tlm_transaction &t);
```

where the parameter t is a reference to the transaction structure and the return value provides status information (i.e. transaction completed successfully or not and maybe further details in case of failure).

This API can be considered as a generic transport layer. The actual content (i.e. the meaning) of the transaction is considered as another layer to define on top of this API. The mechanism to define this content is as follows: tlm_transaction_data is an empty class. The actual structure being transferred is a class instance whose class derives from tlm_transaction_data:

```
class MyTransactionData : public tlm_transaction_data {
  // List of  attributes of the transaction
  ...
  // List of methods to manipulate the
  // transaction attributes
  ...
};
```

TLM protocol:
Relying on the transport API, it is possible to define transaction structures that reflect the protocols required to exchange data. Structures corresponding to the following TLM protocols are available as examples and also with the aim of covering most needs for modeling SoC on-chip communication mechanisms.

- *TAC* for support of interconnection of IP-A models. Transaction is either a read or a write operation.

- *AMBA* for support of IP-MA models based on the AMBA protocol. Single transfers, bursts, and locked transfers are supported.

- *STBus* for support of IP-MA models based on the STBus protocol. Request/acknowledge and separation between requests and responses are supported, supporting sequences as described in Figure 2.2. Such generic protocol can also be derived for specific ones such as the STBus.

- *OCP* for support of IP-MA models based on the OCP protocol. Features are very similar to the STBus.

Communication protocols that require more than one exchange to transfer data can be decomposed into multi way transfers. As an example, a 2-way transaction, as described in Figure 2.11, with request and response semantics, will enable bus protocols such as STBus and OCP to be modeled accurately. The request is initiated when the port is accessed, i.e. myport.transfer(&trans). The request is acknowledged when the transfer function (on the initiator side) returns. An identical scheme holds for the response. Such a decoupling between requests and responses enables separate pipelining of requests and responses.

2.5   Standard Usage of the SystemC API

We propose to implement the API in SystemC as follows: The meaning of the icons and colors of the figures are indicated in Figure 2.9.

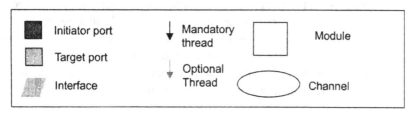

Figure 2.9.   Drawing conventions

An IP is a *module*. An IP includes a *process* if it needs to initiate transactions. Access points are *ports*. A transaction is initiated by transmitting the transaction through an initiator (i.e. master) port. The transaction is conveyed from the initiator module to the target (i.e. slave) port of the servicing module with IMC (Interface Method Call).

Figure 2.10.   Modeling API

The API is basically composed of only one function named

$$transfer(t\_trans),$$

t_trans being the transaction structure described in section 2.4.2.

Life Cycle.   The life cycle of a transaction exchange follows:

1 The process initiates a transaction by calling the port method named transmit. The channel implements the transmit function (i.e. the interface);

2 The transmit function in the channel will then, in turn, activate the service of the module that participates in the transaction exchange by calling the target port method named transfer. The interface of the channel and the target module are identical, therefore enabling a direct connection from the initiator port to the target port with a direct (SystemC) binding of the ports;

3 Transaction service completes when the interface method of the target returns;

4 [optional] A wait statement is executed before the user code starts executing again, therefore providing a mechanism to model concurrency on each transaction 'boundary'. This feature is mandatory for cases when transactions might represent system synchronizations (e.g. polling). When a transaction ends with a call to a wait statement, we name it a scheduled transaction. It is otherwise named unscheduled transaction;

5 The transaction is acknowledged when the initiator port method (transfer) returns.

As depicted in Figure 2.10, the proposed mechanism enables the code of the transaction transfer to be executed in the initiator thread, therefore reducing the total number of required threads (and the related number of context switches). Obviously no restriction is imposed on the receiving module regarding the usage of SystemC processes to model the IP (i.e. the other part of the IP behavior may require the usage of processes).

When timing is annotated on the models, the following timing attributes of a transaction can be identified: initiation time; grant time; reception time; servicing delay; and acknowledgement.

Routing. The address field of the transaction is used to route the request to the relevant module servicing the transaction. Channels can either rely on their own information to route the request or they can use a *decode* method (if provided in the API) to access relevant information located in the targets. The memory map information can either be distributed over the IP models (each model knows its address range) or centralized in the channel. Intermediate solutions can be proposed: information is distributed but cached in the channel at initialization.

Multi-Way Protocols. Protocols that require several transactions to complete an information transfer can also rely on the TLM API. Bus protocols with asynchrony between request and response (e.g. OCP) will rely on a 2-way transaction mechanism as described in Figure 2.11.

Abstraction Adapters. We propose to insert abstraction adapters, as proposed in the SystemC manual, by inserting a module. Figure 2.12 shows an example of an IP-A model (therefore communicating with a 'TAC protocol') adapted to an OCP protocol. All mechanisms rely on the TLM API.

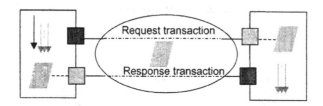

Figure 2.11. OCP-like protocol support

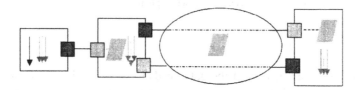

Figure 2.12. Abstraction adapter

2.6 Advanced TLM API Usages

The previous section describes the basic usage of the API to connect IP models to create a SoC simulation platform. The models are connected via a primitive channel that routes the transactions from IP to IP.

Complex communication structures (e.g. cross-bars, layered busses) can be modeled by replacing the primitive channel by a hierarchical channel, as depicted in the simplified Figure 2.13. As shown in this figure, the channel can instantiate ports for its connections. The channel can itself be decomposed of interconnected modules if necessary. Such an approach can be used to model arbitration policies, crossbars, etc.

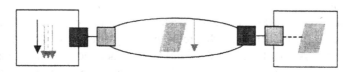

Figure 2.13. Connecting IP models with hierarchical channels

All prior examples describe how to use channels to transport transactions from IP to IP. The transactions are 'serviced' by the IP that participates in the transaction. It is however possible to foresee a usage of the TLM API in which the channel implements the service itself (instead of merely conveying the information), as shown in Figure 2.14. An IP might for example put data in

52

the channel while another will access it. In this case both IPs will initiate the access to the channel.

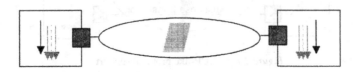

Figure 2.14. Transaction service implemented in the channel

This scenario can be used if the data needs to be stored in the channel (e.g. FIFO, mailboxes). This approach can be suited for example for data flow modeling. In this case, the transaction operations are simple read/writes, as proposed in TLM-A.

By using the same TLM API the mapping process from pure functional models (connected via FIFOs for example) to SoC-A and then refinement to SoC-MA levels will probably be facilitated because automation capabilities can be reused from one level to another.

2.7   Example of a Multimedia Platform

We present in this section a model of a MPEG4 coder/decoder (codec) that uses the API described in the previous section. The need for a model at the SOC architecture level is two-fold: first, an efficient simulation model is required to develop the embedded software and offer interactive debug capabilities; second, early architecture analysis was suitable before the RTL model was ready.

A first experiment had been done in the past to integrate the functional models of the different internal operators, written in C, into a single simulation infrastructure. This infrastructure has been designed to support the concurrent execution of the different threads that represent the partial behavior of the whole system on chip. At that time a basic simulation kernel dedicated to scheduling of the threads was written, using POSIX threads and semaphores. Dedicated to a specific configuration of the system, the main drawback of this C based infrastructure is the restriction in terms of scalability and also the confusion between the models and the simulation kernel itself. Consequently another infrastructure has been developed, using SystemC 2.0 and a user defined channel, that provides the developers with SOC architecture level communication primitives.

In the following we present how this model has been designed, and how it addresses the two needs listed above.

2.7.1   System Model

The platform uses a distributed multi-processing architecture with *2 internal busses and 4 application specific processors (ASIP)*, on which is distributed the embedded software, *a general purpose host processor* managing the application level control, and *7 hardware blocks*. There is consequently a lot of *software parallelism*, with explicit synchronization elements in the software. Each processing block is dedicated to one part of the CODEC algorithm.

The partitioning between Hardware/Software has been done based on the complexity/flexibility needed by the CODEC algorithm. Each operator works at the macro block level (video unit for CODEC). Description of some representative blocks of the system follows.

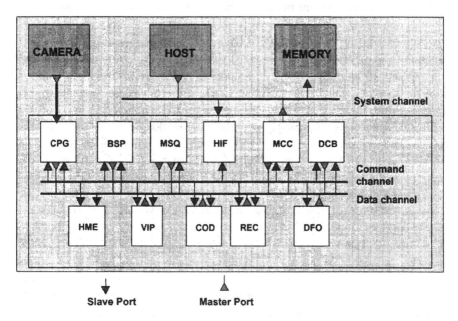

Figure 2.15.   The multimedia platform block diagram

Multi-sequencer MSQ: it manages the video pipeline, is mainly firmware with a hardware scheduler for fast context switch and process management. It is a RISC processor.

Multi-channel controller MCC: hardware with microprogrammed DMA. The MCC arbitrates all the present requests from the operators and performs the I/O with external memory. Scheduling is done per request; a request is one memory burst (variable size). An I/O request is considered as a single transaction even if it takes several cycles to execute.

VLIW image predictor VIP: it is a mix of hardware/firmware: control is done by firmware; processing is done by hardware with a special instruction set of VLIW type. The instruction set is modeled. It performs motion compensation.
Encoder COD: purely hardware, it performs the difference between the grabbed and the predicted image, the DCT, zigzag, quantization and run/level coding.

In terms of system behavior the external processor posts commands to the mailbox (the HIF block). The multisequencer programmed in C in the codec (MSQ) takes the command in consideration, and has in charge the internal control of the codec. Consequently it activates the different internal hardware or programmable blocks to achieve the coding or the decoding of the video flow, by reading status registers and writing command registers for each block. All the operators are pipelined, and communicate with the system memory through a memory controller which is described below. Each internal module has in charge a part of the system behavior, through one or several processes. The memory controller receives requests from the internal operators and generates transactions on the top-level bus to have access to the system memory. It communicates with the internal operators through well identified input/output registers that contain the values to be stored in and loaded from the memory.

In the SystemC model the different blocks are modeled as SystemC modules. Each module has one or several processes (which generally are defined as SC_THREADs), communicates with the other modules through two transaction level communication channels throughout the communication ports. The modules have a role of master when they initiate transaction requests to the memory, and have a role of slave from the memory controller point of view, since this latter generates read and write operations for the I/O registers of the modules. Consequently their communication ports are bound to the data channel for data communication. They also have a communication port bound to the command channel to give access to their control registers. The internal scheduler is a particular case, since it also generates read and write operations to control the system behavior. Consequently it is connected to the command channel through its command master port. Since it also initiate transactions to the memory, a communication port is also bound to the data channel.

2.7.2 Design Choices

The requirement of being able to simulate the embedded software prevents us from developing a purely functional and sequential model of the circuit. Each computing block is modeled as a SystemC module, with its own processes and associated synchronization elements. Three categories of blocks have a specific modeling strategy:

Programmable Modules. This multi-processor design is composed of C programmed modules as well as pure hardware blocks and mixed blocks that encompass hardware and programmable operators.

Software block: we rely on the sequential aspect of the code, for the MSQ, BSP, HME, and VIP. All the firmware is written in C. This enables the code to be natively compiled on the workstation and to communicate with the TLM model through an I/O library (via simple stubs to call C++ from C). The specific built in instructions of the processors are modeled as C functions. One part of the C model is used directly for generation of ROM code, using retargetable C cross-compilers. An instruction set simulator (ISS) has been integrated in the environment to run the cross-compiled application software, running on the SoC host. Hence the software is written once, and remains unchanged whatever is the environment: TLM and RTL simulations, emulation, and application board.

Hardware block: a high level model is written in SystemC. It is a functional, bit true model with the representation of the transactions with memory. We do not represent the internal structure of the block. We model the input/output of the block, the synchronization and the internal computation at the functional level. The complete SystemC model is used as reference during RTL validation.

Let us take an example. The coder block (COD) is a hardware pipeline with 5 operations: Delta, DCT, Zigzag, Quantization, and run/level. A FIFO is inserted between all these operations; the computation, controlled by an FSM, is pipelined. A DMA manages the inputs and the outputs. The RTL model is fully representative of this architecture. Such a model requires at least 3 men months effort for a senior designer. In contrast, the corresponding transactional model will consist in input data acquisition (grabbed and predicted blocks in our example), result computation by a C function, and result output. The gain is two fold: writing models is easier and faster, and simulation speed is greater. The corresponding effort is about 1 man.week.

Mixed hardware/software block: mixed hardware/software blocks are modeled as a mix of the two previous categories.

2.7.3 System Integration Strategy

In this part we illustrate some features of the modeling environment used in this platform. Mechanisms such as concurrency, hierarchical modeling, source level debug and monitoring, natively supported by the underlying simulation technology, are not detailed here. Rather we focus our attention on the fundamental aspects of our approach.

System synchronization in the platform is two fold. First, the MSQ has in charge the global control of the Platform. It is responsible for the activation of the coding and decoding tasks according to the current status of the system. Hence a task will be executed only if the relevant data is available, and will be suspended otherwise. Because of the internal pipeline of operations several tasks may be activated at the same time. Synchronization is achieved by writing/reading command and status registers of the different operators. Second, data exchanges between the operators and the memory are blocking operations. The synchronization scheme of the platform ensures that an operator will resume its computation only once the previous transaction is completed from the system point of view.

In the platform we model the data exchanges with arbitrary sizes, with respect to the system semantics of the exchange. Transactions between the camera and the grabber transfer images line per line, whilst transactions between the MSQ and the other operators are 32 bits wide.

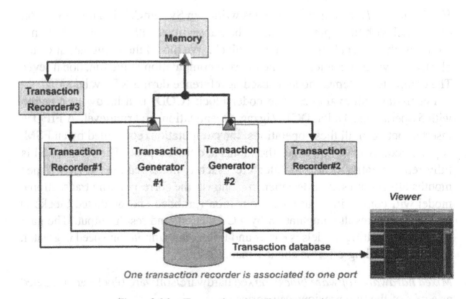

Figure 2.16.   Transaction recording mechanism

2.7.4   Monitoring

Monitoring at the SOC architecture level is well suited to assisting developers in the understanding of the data exchanges in the system. It helps for embedded software debugging (understanding write and read operations on command/status registers), and it is mandatory for understanding the temporal behavior of the different concurrent processes which are executed in the different modules. In order to avoid a heavy instrumentation of all the models, which is tedious and error-prone, we have implemented a generic transaction recording mechanism, which is instantiated in each master and slave port, in order to record all outgoing and ingoing transactions (see Figure 2.16).

For ease of use and simulation efficiency it may be enabled and disabled on a port by port basis. Each port may instantiate a transaction recorder that dumps the transactions into a database. This database is then loaded in a wave viewer, and queries may be executed to select a specific subset of the exchanged transactions, and visualize specific attributes (data, address, duration, etc).

2.7.5   Experimental Results — Simulation Figures

We compare the performance figures obtained for this multimedia platform in terms of code size and execution speed on different complementary environments in Figure 2.17.

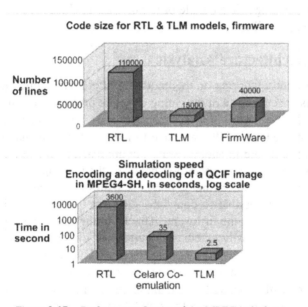

Figure 2.17.   Performance figures of the MPEG4 platform

The modeling choices inside the IPs enable a significant gain in terms of model size: TLM models are 10× smaller than RTL. Consequently they are easy to write and fast to simulate. The simulation speed is 1400× faster for TLM compared to the RTL model on a SUN Ultra 10 workstation/330 MHz, 256 MB memory.

The RTL model simulates one coded image in one hour, whilst TLM models in SystemC require 2.5s per coded image. Before the availability of the Accellera SCE-MI interface [Accellera, 2002a] on the Mentor Celaro hardware emulator, an *ad hoc* co-emulation transactional interface has been implemented thanks to the C API that provides both controllability and observability of the emulated design (clock control, memory and register load and dump, programmable built in logic analyzer). On the hardware side synthesizable models have been developed for the camera, the memory, and several RTL transactors [Accellera, 2002a] that translate a data packet from a transaction into a bus cycle accurate data exchange. On the software side the host TLM model has been extended for HW debugging and performance evaluation purposes. The co-emulation requires 35s per coded image, i.e. a system clock at about 40–60 kHz. This speed (more than 30 times faster than a cycle based co-emulation) enables running the software developed onto the SoC TLM platform without any external or synthesizable CPU core aimed at the host modeling.

The performances obtained for the transaction level modeling environment are well suited for embedded software development. Since it runs a significant test bench of 50 images in a couple of minutes on an Ultra 10, with full source level debugging facilities.

2.7.6   Architecture Analysis

We are currently working on this model to offer architecture analysis also. A first step is the capability of providing dynamic profiling information for the embedded software execution. Since instruction set simulators are not available for the four embedded ASIPs, we have to identify the computational cost of the embedded software in another way. An approach is to overload the usual arithmetic and comparison operators thanks to the C++ operator overloading mechanism, to add counters that generate statistics about the execution of the firmware. So it is possible to count the different C-level operations ($+$, $-$, $<$, $=$, etc.) executed in the firmware. Thanks to the specificities of the ASIP architecture one can easily associate a number of cycles to each source level operation. Consequently, it is possible to add delays in the execution of the firmware before each transaction on the communication channel, to model the duration of the embedded software computation. In this case the computation is still executed in zero time from the simulation time point of view, but a delay is then inserted to delay the generation of the next transaction.

The timing of the hardware blocks is achieved by back annotation from the RTL models when available, or according to the designer's expertise, as discussed in subsection 2.3.6.

2.8   Example of ARM Processor Subsystem

This section discusses the necessary modeling decisions and associated tradeoffs when developing TLM models and how they impact subsequent activities such as performance analysis.

We explain the abstraction choices made while modeling the ARM PrimeCell Direct Memory Access Controller (DMAC PL080) of the PrimeXsys Wireless Platform [ARM, 2002]. This DMA controller is as a good example of TLM modeling choices because the controller needs a parallel style of modeling to capture correctly the whole behavior. Also some of this hardware parallelism is not controlled by software but is hardware logic only.

2.8.1   DMAC PL080 TLM Model

The main features of the ARM PL080 DMA Controller are the following:

- 8 prioritized dma channels;

- Word, half word, and byte transfers;

- 2 master AHB bus ports, allowing more data throughput on 2 busses;

- 16 peripheral control interfaces. Such an interface allows transfers to be controlled by the peripheral instead of the DMA controller;

- Both little endian and big endian transfer modes. The 2 AHB ports can be separately programmed to support each mode;

- Peripheral control interfaces can be controlled either by hardware signals or by using bits in special registers called SoftXReq that can be set by embedded software external to the DMA;

- Supports an AHB slave interface to program the DMAC internal registers. The registers are memory mapped.

So from the specifications we can see that the interfaces to be modeled in the DMAC TLM model are:

- One AHB slave port;

- Two AHB master ports;

- Sixteen Peripheral control signals. DMACSREQ, DMACBREQ, DMACLBREQ, DMACLSREQ, DMACCLR, DMACTC;

- Three DMA Interrupt signals. DMACINTERROR, DMACINTTC, DMACINTCOMBINE.

Based on the understanding of the above features we defined the TLM model interfaces and internal behavior.

Data ports such as AHB and APB will rely on the TLM API. System synchronizations will rely on the native simulator mechanisms. In SystemC, either signals or events are available. It is recommended to use signals to ease the usage of the TLM model as a golden reference model for the RTL verification and also for co-simulation activities.

AHB slave Port: This interface is used to access the DMAC registers. One should take care that the (*a priori* asynchronous) update of DMAC TLM slave registers by the external embedded software should not corrupt the internal behavior that relies on them. SystemC runs with non-pre-emptable threads. The systemC simulation can be controlled in each thread to yield control only at known pre-emption points. It is the responsibility on the designer of the module to ensure that pre-emption points are carefully defined. Modeling the system level synchronizations helps to achieve both a safe system design and a correspondingly correct SystemC simulation.

AHB Master Ports: The real DMAC has two master ports, allowing two simultaneous read and write operations. In a TLM platform used for architecture analysis, both ports are required in order to provide realistic traffic generation. Even in a TLM platform used only for embedded software development there is a need for modeling the two ports because each one has specific registers that are accessed by the embedded software.

Three Interrupt Signals: They represent system events and therefore need to be modeled.

Register bank: The registers are modeled as data members of TLM modules class definition. For DMAC TLM the size of all the registers can be made equal, e.g. to 32 bits (the maximum register size or host machine word size, whichever is maximum).

As a general rule for defining how to model the internal TLM behavior of an IP, we can take any deviation from functionality of the block as far as the functionality of the block remains the same from the software point of view. These abstractions help reduce the modeling effort and simulation time of the block.

The DMA includes two internal arbiters (one per bus port). The arbiter implements a priority-based algorithm, i.e. the highest priority channel with pending transactions will be allowed to transfer data. When a channel suspends transferring data because no input data is available, the arbiter will grant access to another channel of lower priority. This last feature enables the optimization of data transfers. It does not need to be part of the TLM model to ensure the correct execution of the embedded software. However, this model cannot be simply annotated with timing information to reflect its correct timing behavior. A correct interleaving of data transfers between active channels must be modeled in this case. The addition of such scheduling scheme will refine the model into a TLM-MA level suited for architecture analysis. The model can be compiled with or without the added functionality.

2.8.2   ARM Subsystem Performance Analysis

Let us describe results of using the DMA TLM with the back-annotated TLM model of the static memory controller, as part of the PrimeXSys TLM ARM subsystem simulation that is close to RTL precision — yet available before RTL and simulating orders of magnitude faster.

The TLM platform for embedded software needs to be upgraded for enabling performance analysis. The main reasons are:

- Traffic generated by the processor on the busses highly depends on its cache behavior;

- Multi-layer bus structure includes some arbitration scheme to access shared slaves;

- Multi-port memory controllers include arbitration policies to select incoming requests;

- Memory access wait states is not a fixed value.

In order to build a suitable platform for performance analysis, the following steps are performed:

- An instruction set simulator (ISS) which includes a model of the cache for the processor is integrated in the platform.

- The initial DMA TLM model is refined into a TLM-MA model, as described in the previous paragraph.

- Realistic arbitration policy that controls accesses to shared resources is included.

Memory latencies are accurately modeled, as described hereafter. Our analysis focused on the static memory controller. The number of wait states observed at a given port depends on current and previous transfers of accesses on that port. If an access is initiated when another transaction is accessing the static memory, the second transaction finishes a fixed time after the first one. This duration depends (only) on the memory area being accessed (which bank), and on the access directions (read/write) of the previous and current transactions. For a 4-port, 8-bank memory controller there are 256 situations, each one with a fixed number of wait states observed in RTL simulation. As a result the way to give an annotation by time for the TLM of this static memory controller is to overload the read/write methods of the functional TLM model with read/write methods that call the SystemC wait function with a delay equivalent to the given number of cycles for that situation (this duration is one of the 256 values). The timed TLM simulation of the memory controller provides the adequate cycle count estimation compared to RTL simulation.

The resulting simulation platform comprises a mix of TLM-A and TLM-MA models with sufficient timing accuracy to be used for performance analysis.

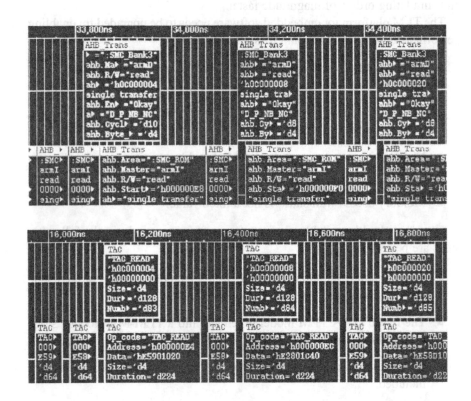

Figure 2.18. Traces for performance analysis

Such a trace can be viewed in Figure 2.18. The upper diagram is RTL and lower diagram is TLM.

In each diagram the upper line represents transactions on the instruction bus (AHB-I), the lower line displays transactions of the data bus (AHB-D). Both busses are connected to a single Static Memory Controller (SMC) with separate ports. Data is stored in a flash while instructions are stored in a ROM. Hence the two AHB busses are concurrently trying to gain access to memory banks via the SMC. When such conflicts occur, the instruction fetch transaction takes longer.

This is visible on both the RTL and the timed TLM simulation. In fact, thanks to back-annotation from RTL of the memory controller the accesses have the same duration as in the RTL simulation while running at the TLM simulation speed. As a conclusion, with time annotated TLM models, architects can now estimate SoC performance with light modeling efforts and fast simulations which include both hardware and software, with easily modifiable hardware models, without having the RTL of the whole SoC, and without modeling at cycle accurate (low) level.

2.9   Conclusions

As highlighted in the chapter, we have introduced a new abstraction level in our SoC design flow to address the needs of early embedded software development, architecture analysis, and functional verification. This level captures the architecture of the IP (or SoC) and complements the RTL level which models the microarchitecture. The combined usage of these two views avoids the usual conflict between the need for accurate descriptions and the request to simulate at high speed the real time embedded application software. This flow extension has successfully been applied to SoC design and verification activities in an industrial context involving very large teams with restricted manpower dedicated to the introduction of new techniques.

Acknowledgments

This chapter is an overview of the system level design flow developed at STMicroelectronics. We would like to thank all our colleagues who contributed to the developments described in the chapter. In particular, we thank Eric Auger and Nicolas Mareau for their major contributions to the development of 'RTL platforms', Antoine Perrin and Sudhanshu Chadha for the development of SysProbe and Rohit Jindal for his contribution to TLM models. We also gratefully acknowledge the support of Srikant Modugula for the coordination between the teams of Crolles (France) and Noida (India). Thanks to Jean-Claude Bauer and Murielle Icard-Paret for their active contribution to the review and improvement of the chapter. Last, but not least, we would like to thank Philippe Magarshack for his continuous support and encouragement during the development of the flow.

Chapter 3

Refining a High Level SystemC Model

Bernhard Niemann, Frank Mayer,
Francisco Javier, Rabano Rubio, Martin Speitel
Fraunhofer Institute for Integrated Circuits, Erlangen, Germany

Abstract   The objective of this paper is to present a possible flow, using SystemC, to make the transition from a high level data flow description towards an implementable model. The main focus is behavior refinement, in other words the modification of the module descriptions until they can be mapped to a given architecture, satisfying the implementation constraints. A customizable processor core that executes an ANSI C version of the system model is utilized. Only operations that require a lot of computational power are implemented as custom hardware modules to extend the core functionality. A simple operator counting technique is used inside the high level model to obtain run time estimates for hardware–software partitioning.

3.1   Introduction and Motivation

Introduction.   With SystemC 2.0 a C++ class library is available which claims to serve as a unified modeling and implementation environment for software as well as hardware. SoC development using SystemC starts with a high level model of the system which exhibits the intended behavior when executed (the so called *Executable Specification*). This model is then partitioned into hardware and software, and mapped to a target architecture (*Transaction Level Model* or *Architecture Model*). The hardware modules, the software modules, and the communication protocols of the architecture model are then gradually refined towards an implementable model, always using the high level model as a golden reference.

Motivation.   The motivation for the work presented here is to explore a possible design flow using SystemC for rapid prototyping and implementation of a software Orthogonal Frequency Division Multiplex (OFDM) demodulator. As much processing as possible should be performed in software running on

W. Müller et al. (eds.),
SystemC: Methodologies and Applications, 65–95.
© 2003 *Kluwer Academic Publishers.*

a customizable processor core. Functions which are computationally intensive can be either implemented as additional hardware modules or by using custom operations that extend the instruction set of the processor core.

SystemC is used to implement the high level reference model and to easily retrieve first performance estimates of an implementation on the target architecture. The high level model is then refined to ANSI C for several reasons. First, to have software running on the target system as early as possible; second, to verify the performance estimates obtained from the high level model; and third, to serve as a common basis for possible further refinement to various different target architectures.

As a pure software solution is too slow, the run time estimates obtained in previous steps are used to partition the system into hardware and software components. The implementation of custom operations to speed up the C program is demonstrated using Fast Fourier Transform (FFT) butterfly operations as an example. SystemC is used to explore and verify the implementation of various possible custom operations.

Outline.   This article consists of four main parts. The first two sections deal with the description of the OFDM system and the system model used for this work. After a brief presentation of the OFDM demodulator and its basic concepts a thorough description of the high level data flow model is given. The principles of data flow modeling and the techniques used to allow for easy refinement in later stages are presented.

The refinement process is presented in the last two sections. They consist of a description of the technique used to get from the data flow model to a first ANSI C model and a discussion of the refinement process used to increase the speed of that ANSI C model. The steps carried out to replace the modules and 'First In First Out' (FIFO) channels of the data flow model by C language constructs are described. Going one step back to the SystemC data flow model, the flow from early performance analysis to implementation of the operator grouping is described.

3.2   The OFDM Demodulator

This introductory section explains the principles of OFDM modulation and the general structure of the demodulator system used in the following sections. We concentrate on a discussion of the most important concepts. For a more thorough treatment of OFDM see, for example, [van Nee and Prasad, 2000].

3.2.1   OFDM Basics

Modern applications for wireless communication and broadcasting systems require high data rates. Additionally, the systems have to be designed for mo-

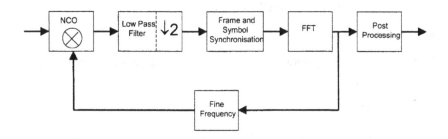

Figure 3.1.   Block diagram of an OFDM demodulator

bile reception and difficult environments with rapidly changing channel conditions. Examples for such systems are the wireless LAN standard 802.11a, or broadcasting systems such as XM Radio [XM Satellite Radio, 2003] and Sirius Satellite Radio [Sirius Satellite Radio, 2003] in the U.S.A.

A common problem of these channels is deep narrow band fade owed to multipath reception in an urban environment and narrow band interference. The basic idea of an OFDM system is to divide the available frequency band into small sub-bands. The distribution of the information over the available subcarriers results in a rather low datarate for each subcarrier. Every sub-band or subcarrier is modulated using Quadrature Amplitude Modulation (QAM). The QAM scheme depends on the signal to noise ratio in the subcarrier, and typically varies from Binary Phase Shift Keying (BPSK) up to 64 QAM.

To have the best spectrum efficiency the different modulated subcarriers have to be orthogonal. By using the Inverse Fast Fourier Transform (IFFT) for modulation the subcarriers are implicitly chosen to be orthogonal. Using an FFT in the receiver, the information carried by the OFDM signal is efficiently restored.

The critical part of an OFDM demodulation is the synchronization of the incoming signal. The FFT processing must be synchronized to the boundaries of the OFDM symbols. The system is sensitive against phase noise and frequency offset. Therefore symbol tracking and the correction of the frequency offset and deviation is crucial.

3.2.2   The Demodulator

The OFDM demodulator as used in the following sections of this article is briefly presented here. The steps that are carried out by the receiver are

- mixing to base band, down sampling and filtering of the received signal;

- fine frequency synchronization (frequency offset correction);

- synchronization of the OFDM symbols;

- subcarrier separation by FFT computation;

- QAM demapping, error phase offset compensation and Viterbi metrics generation;

- error correction.

Figure 3.1 shows a block diagram of the system. The Numerical Controlled Oscillator (NCO) at the input is used for frequency and phase correction and has to shift the required frequency band exactly into the base band. It is controlled by the fine frequency module, which determines the correction signals to the NCO. The discriminator characteristics are implemented in the fine frequency module. After the NCO the incoming data samples are filtered and down-sampled by a factor of two.

The next module is the frame and symbol synchronization, in which the boundaries of each OFDM symbol are detected and the data blocks for the FFT are defined. Within the FFT module the subcarriers of the OFDM symbol are separated. The outputs of the FFT module are complex values for each sub-carrier, representing the transmitted signal information. The post-processing demaps the OFDM symbols, compensates the error phase offset, and generates the Viterbi metrics. The last processing stage is comprised of error correction, interleaving, and de-multiplexing of the data-stream, which are not shown in this article.

3.3 High Level SystemC Model

The following section discusses the high level model of the OFDM demodulator. This model is used as a basis for refinement in the following sections. It is implemented as a so called data flow model using SystemC.

SystemC 2.0 provides the basis for modeling with a variety of different Models Of Computation (MOC). A model of computation is characterized by (see also [Grötker et al., 2002]):

- the model of time;

- the rules for process activation;

- the communication semantics.

Though providing very flexible and powerful means for implementing different MOCs, there are only a few MOCs supported by the pre-defined channels of

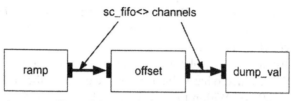

Figure 3.2. A simple data flow system

SystemC 2.0. One of the supported MOCs is data flow modeling, as used for the implementation of the high level model.

3.3.1 Introduction to Data Flow Modeling

This paragraph gives a short introduction to data flow modeling with SystemC. After a brief discussion of the basic ideas of data flow modeling a simple example using SystemC will be discussed.

Data flow modeling is especially suitable for describing the data path of an algorithmic system as found for example in wireless, telecom, or broadcasting applications. The different algorithms comprising the system are described in separate modules (functional decomposition). Each module has one or more ports. Input ports are used for reading data, output ports are used for writing data. The modules are connected to each other using FIFO channels bound to the ports of the modules.

Communication between the modules is handled by blocking read and write operations. A blocking read on a FIFO channel only returns when a sample is available in the FIFO. If the FIFO is empty execution of the module will block until at least one sample is written to the FIFO. On the other hand, a blocking write on a full FIFO channel will stall execution of the module until at least one sample has been read from the FIFO.

With SystemC 2.0, implementing a data flow model is relatively simple and convenient, as the 'core language' standard already contains all the language constructs needed to describe modules with FIFO ports that are connected by FIFO channels. See table 3.1 for an overview of the mapping of data flow semantics to SystemC 2.0 syntax.

Figure 3.2 shows a simple example of a data flow system in SystemC. The example consists of a source (`ramp` module), a very simple algorithm (`offset` module) and a sink (`dump_val` module). All modules are connected by channels of type `sc_fifo<>`.

The code of the `offset` module is shown below as an example. It is implemented as a 'template class' in order to be able to use the module with different data types. At module instantiation a template parameter determines the data type of the samples.

The module has an input port inp of type sc_fifo_in<T> and an output port outp of type sc_fifo_out<T>. Both ports are public members of the module, because they need to be accessed from outside the module at instantiation, in order to connect FIFO channels.

As the constructor should take an argument to determine the offset value, the SC_HAS_PROCESS macro has to be used before defining the code of the constructor [OSCI, 2002a]. Within the constructor the member function main() is registered as an SC_THREAD process. An empty destructor is given for completeness but could have also been omitted in this simple example.

```
template< class T > class offset : public sc_module
{
public:
  sc_fifo_in< T > inp;
  sc_fifo_out< T > outp;
  // use SC_HAS_PROCESS together with custom constructors
  SC_HAS_PROCESS(offset);
  offset(sc_module_name nm, int off_val = 0)
    : sc_module(nm), _off_val(off_val)
  {
    SC_THREAD(main);
  }
  // destructor
  ~offset() {}
  // member function registered as process
  void main();
protected:
  T _off_val;
};
```

Semantics	SystemC 2.0 Syntax
Module	class my_module : public sc_module{};
Algorithm	SC_THREAD(main);
FIFO Input Port	sc_fifo_in<data_type> input_port_nm;
FIFO Output Port	sc_fifo_out<data_type> output_port_nm;
FIFO Channel	sc_fifo<data_type> fifo_nm("fifo_nm", size);
blocking read	input_port_nm.read();
blocking write	output_port_nm.write();

Table 3.1. Mapping of data flow modeling semantics to SystemC 2.0 syntax

The algorithm of the offset module is implemented inside the member function main() that is registered as an SC_THREAD process. The code for the process has to appear in the header file along with the class declaration because offset is a template class. To avoid linker errors resulting from symbols that are defined more than once, main() has to be specified as in-lined member function.

```
template< class T > inline void offset<T>::main()
{
  T tmp;
  // infinite loop used inside SC_THREAD process
  while(1) {
    tmp = inp.read(); // blocking read
    tmp += _off_val;   // the algorithm
    outp.write( tmp ); // blocking write
  }
}
```

The process code follows a scheme used for all the modules of the high level model. First of all, note that the process code is enclosed by an infinite loop, which is required by the SC_THREAD process type [OSCI, 2002b] [Niemann, 2001]. Inside this loop the first action is to read the value of the input port to a temporary variable using blocking read. The temporary variable tmp is defined outside the infinite loop to avoid the run time overhead of re-defining the variable at each loop iteration. The next line represents the data processing of the module, which is performed using the temporary variable (of course, in more complex modules, data processing will require much more than one simple line of code). After all processing has been done, the result is written to the output port using a blocking write operation.

3.3.2 The OFDM Demodulator Model

The OFDM demodulator, as described in section 3.2, has been implemented as a data flow model using SystemC 2.0. A schematic of the model is shown in figure 3.3. The schematic has been obtained using a screen shot of the model imported into Synopsys CoCentric System Studio™. At the time the high level model was implemented, the SystemC integration of System Studio was not available, and most of the work presented in this article has been carried out using the free reference implementation of SystemC.

Module Implementation.   It can be seen that every algorithmic task is implemented in a separate module. See table 3.2 for an overview of the modules, their tasks in the overall system and their I/O behavior. The inputs to the

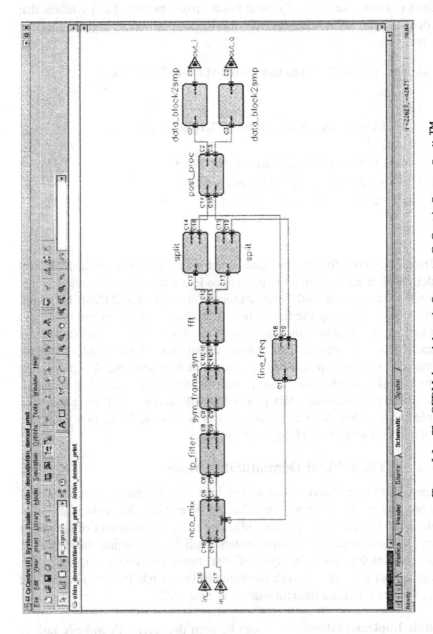

Figure 3.3. The OFDM demodulator in Synopsys CoCentric System Studio™

module name	function	I/O: block or sample based
nco_mix	complex mixer (remove freq. offset)	sample
lp_filter	filtering and down sampling by two	sample
sym_frame_syn	symbol and frame synchronization	input: sample output: block
fft	fast Fourier transform	block
post_proc	demodulation	block
fine_freq	discriminator and loop-filter	input: block output: sample

Table 3.2. Overview of the algorithmic modules in the OFDM demodulator model

high level model are two streams of floating point samples interpreted as one stream of complex floating point samples. After the samples have passed the nco_mix and the lp_filter modules, synchronization is performed by the sym_frame_syn and the samples are aligned and put together to data blocks. All the further processing (fft, post_proc and fine_freq) is block based.

The OFDM demodulator contains one feedback loop for frequency control. A feedback loop in a pure FIFO based data flow model would be implemented using a delay module inside the feedback loop. For the OFDM demodulator, the delay module would have to be placed between fine_freq and nco_mix, because otherwise the nco_mix module would try to read a sample from an empty FIFO in the feedback loop and the complete simulation would be blocked. The approach used in the model presented here is to use a signal instead of a FIFO for the connection of fine_freq and nco_mix. This has the advantage that the frequency correction is applied to the nco_mix module as soon as it is available inside the fine_freq module. Moreover, as reading from a signal never blocks, a delay module is not necessary in this case.

To enable a smooth refinement some simple rules were followed during the implementation of the high level model:

- Use typedefs for the data types of all input and output ports;

- Only use one process within each module;

- Structure the algorithm into different sub-tasks and use member functions to encapsulate those tasks.

Apart from the six algorithmic blocks, there are two modules which control the data flow through the system (split and data_block2smp). Table 3.3 lists the properties of these modules. The data_block2smp module is used to

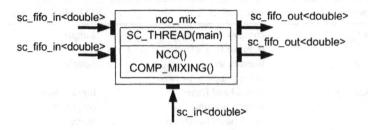

Figure 3.4. The complex mixer module

convert blocks of data back to a stream of samples, which can then be written to a file. The split module is necessary because in SystemC an sc_fifo<> channel may only be connected to one input port. However, the output of the FFT is needed in the post_processing module as well as in the feedback loop to calculate the frequency offset estimates for the NCO. The splitter modules simply copy data of one input port to multiple output ports (in this case two). The split modules as implemented in this model can handle any data type that has an assignment operator (operator=()) defined.

module name	function	I/O: block or sample based
split	duplicate the number of streams	sample or block
data_block2smp	convert blocks to consecutive samples	input: block output: sample

Table 3.3. Overview of the modules in the OFDM demodulator model used for data flow control

The Complex Mixer as an Example. To give a concrete example of the implementation of a module, the complex mixer module (nco_mix) will be discussed in some detail. The general structure of the module is depicted in figure3.4. It has two FIFO input ports and two FIFO output ports, one of each for real and one for imaginary data. The additional input for the frequency is implemented as a signal port (see page 73). The SC_THREAD process main() runs inside the module and calls the functions NCO() and COMP_MIXING() that perform the actual algorithm.

The source code for the declaration of the nco_mix module is given below.

```
#include "ofdm_demod_pkg.h"
class nco_mix : public sc_module
{
```

```
public:
  sc_fifo_in<T_FFT>   in_data_r;
  sc_fifo_in<T_FFT>   in_data_i;
  sc_in<T_FREQ>       in_nco_inc;
  sc_fifo_out<T_FFT>  out_data_r;
  sc_fifo_out<T_FFT>  out_data_i;
  SC_CTOR(nco_mix) {
    SC_THREAD(main);
  }
  // process prototype
  void main();
  // function prototypes
  int NCO(const T_NCOINC& incr,
          T_FFT& cos_val, T_FFT& sin_val);
  int COMP_MIXING(T_FFT& in_r, T_FFT& in_i,
                  T_FFT& out_r, T_FFT& out_i,
                  T_FFT& cos_val, T_FFT& sin_val);
private:
#ifdef DEBUG
  ofstream _ofs;
#endif //DEBUG
};
```

It can be seen that the user defined types T_FFT and T_FREQ are used for the ports of the module. Those types are defined inside the file of dm_demod_pkg.h which is included by all modules. The following code snippet, for example, sets both types to float

```
typedef float T_FFT;
typedef float T_FREQ;
```

Inside the constructor the method main() is registered as an SC_THREAD process. The class declaration only contains the prototypes for the process and the other functions. The definition is given in the implementation file (nco_mix.cc) that is not shown here. At the end of the class declaration it can be seen that some additional code for debug purposes has been added.

Communication.   As most of the processing in the OFDM demodulator is block based it is convenient to create a data_block class that is exchanged over the FIFOs. The advantage is that only one read or write access to the FIFO is necessary to exchange a complete block of samples. The data_block class is implemented as a template in order to allow blocks of different data types. It is important to note that several operators and functions need to be implemented for a user defined type if it should be exchanged over a SystemC 2.0 sc_fifo<> channel [OSCI, 2002a]:

- default constructor
  T()

- assignment operator
  operator=(const T&)

- output stream (stream insertion) operator
  operator<<(ostream&, const T&)

Parts of the declaration of the data_block class are shown in the following code sample.

```
template<class T> class data_block
{
public:
  // constructors
  data_block(size_t len = 0);
  data_block(const data_block& rhs);
  // destructor
  ~data_block();
  // ...  other methods not shown ...
  // operators
  bool operator==(const data_block<T>& rhs) const;
  data_block<T>& operator=(const data_block<T>& rhs);
  // I/O
  ostream& print(ostream& os) const;
private:
  size_t _len;
  T* _ptr;
};
template<class T>
ostream& operator<<(ostream& os, const data_block<T>& out);
```

The default constructor, needed to pass a data_block object over a FIFO channel, creates a data_block object of size zero. On an assignment of a block of size S to a block with zero size, the block length of the zero size block is extended to size S. This is handled by the copy constructor and the assignment operator, both of which create a deep copy (see [Carroll and Ellis, 1995]). The equality operator compares each element of the left hand side with each element of the right hand side and only returns true if all elements match.

The switch from sample based to block based processing requires some consideration regarding the appropriate FIFO size. In SystemC 2.0 the size of the FIFOs is determined by the user and not by the simulation kernel. If the sizes are chosen inappropriately this will result in unnecessary context switches. A context switch occurs whenever a module tries to read a sample from an empty FIFO or a module tries to write a sample to a full FIFO, in other words, whenever the module is blocked. In the sym_frame_syn module, at least the number of samples comprising one complete block of data is read before the block is written to the output port. To be able to read all samples required for one complete data block from the input without a context switch, the connected FIFOs need to have a size of at least the block size.

Therefore, the following sizing rule was applied to the demodulator model. Let S be the size of one data block and N the size of the FIFOs for block based communication,

$$size\big(\text{sc\_fifo<data\_block<T\_FFT> >}\big) \;=\; N, \tag{3.1}$$
$$size\big(\text{sc\_fifo<T\_FFT>}\big) \;=\; S * N. \tag{3.2}$$

This sizing rule ensures that only after the processing of N complete blocks of data a context switch occurs. The performance gain — in terms of simulation time — resulting from correct FIFO sizing was up to 50% for the OFDM demodulator, compared to worst case sizing (all FIFOs have size one).

3.3.3   Counting the Operations

A simple technique using operator overloading was employed to extract the number of operations that have to be computed within each module. The results can be used to get a first estimate on the algorithmic complexity of each module. The idea behind this is to provide preliminary performance data as early as possible in the design flow. Based on this data, the computational load of various algorithms may be compared, and moreover, the data can help to make a decision on the partitioning in hardware and software. This will be discussed in more detail in section 3.4.4.

In this section, the implementation used to retrieve the operator data is explained. It is based on two prerequisites:

- the modules make use of typedefs to be able to separate the operations in the data path from control structure like branches or loops;

- SystemC provides a mechanism to retrieve the name of the current module and process.

Given these prerequisites, the idea is to replace the data type used in the data path (double in our model) by a user defined template class specialized by the data type (profile< double >). This profiling class provides overloaded implementations of all operators used in the data path. These implementations call the operators of the underlying data type and additionally dump the type of operation, the name of the module, and the name of the process to a file.

The implementation of the operator+= is shown in the following code snippet. It can be seen that apart from performing the add operation, a function pout() is called with the name of the operator (op+=). This function performs the dumping of the necessary information to a file.

```
template<class T>
inline
profile<T>& profile<T>::operator+=(const profile<T>& oper1)
{
  pout("op+=");
  // m_var is a member variable of profile<T>
  m_var += oper1.get_val();
  return *this;
}
```

The function pout() checks whether it is called outside the context of a process (p_hdl->kind == SC_NO_PROC). If this is the case the string no_proc is printed along with the operator type. If the operator was called in the context of a process, the module name and the process name are printed as the context of the operator.

```
// function to write the operator type.
inline void pout(char *string)
{
  sc_curr_proc_handle p_hdl =
    sc_get_curr_simcontext()->get_curr_proc_info();
  if( p_hdl->kind == SC_NO_PROC_ ) {
    ofs_prof << "no_proc";
  }
  else {
    ofs_prof << p_hdl->process_handle->name();
  }
  ofs_prof << " " << string << endl;
}
```

The output of a profiling run for the low pass filter module is shown below. Of course, this data has to be post-processed in order to obtain the number of

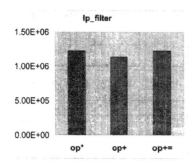

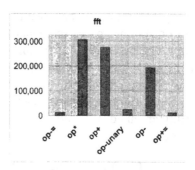

Figure 3.5. Operator histogram of the low pass filter module for one frame

Figure 3.6. Operator histogram of the FFT module for one frame

operator calls for each type of operator. This is done by a simple Perl script that takes the output of a profiling run and converts it into a table containing the type of the operator and the number of invocations. This file may then, for example, be used as input to a spreadsheet program to create a graphical representation of the data obtained.

```
. . .
i_lp_filter.main op+
i_lp_filter.main op*
i_lp_filter.main op+=
i_lp_filter.main op+
. . .
```

Histograms for all modules can be generated which show how often an operator is called for a given module. Two such tables are shown in figure 3.5 for the `lp_filter` module and in figure 3.6 for the `fft` module. The number of processed samples is the same for both figures and corresponds to one frame. A frame is comprised of several OFDM symbols and has a preamble associated with it. The details of the frame structure are beyond the scope of this article.

3.4 Refinement to ANSI C

This section describes refinement of the high level model to an ANSI C program.

3.4.1 Introduction

The ANSI C description can be seen as an intermediate step in the refinement process towards a real time implementation. It may be used as a common starting point for implementation on various platforms consisting of a microprocessor and some custom hardware. The advantage of this model is that it allows for early integration on the target platform and avoids the run time overhead of a C++ or SystemC based implementation.

As will be shown in 3.4.2, the functions performing the algorithms, and therefore the number of operations executed by the model, remain the same during this refinement step. At the end of this section, in 3.4.4, a comparison is made between the run time of the model on the target processor, an ARCtangent[TM]-A4 (further called A4), and the estimates obtained from the operator counting technique.

3.4.2 Replacing the Modules

The objective for the refinement to ANSI C is to obtain a running system as early in the design process as possible. Therefore a straightforward solution, based on the usage of a framework, was used for the replacement of the modules.

The idea is to substitute modules by function calls. A simple framework handles the calling sequence of the functions and thereby replaces the SystemC scheduler and the implicit synchronization offered by the `sc_fifo<>` channels. Moreover, it handles the input and output buffers of the functions as a replacement for the storage associated with the `sc_fifo<>` channels.

SystemC Module	ANSI C
SC_MODULE	functions called by a framework
member variables	`static` variables in `<modul_name>.c`
input and output ports	pointers to input and output buffers
constructor	use special initialization phase
SC_THREAD	no process registration necessary
local variables of SC_THREAD process	`static` variables of corresponding function
`read()` and `write()` methods	pointer access to buffers
implicit synchronization with FIFO channels	finite state machine of framework

Table 3.4. Mapping of SystemC language elements to ANSI C

The mapping of SystemC language elements to ANSI C is summarized in table 3.4. As an example to illustrate the mapping process, the complex mixer introduced on page 74 will be discussed in some detail.

It can be seen that the class declaration for the `nco_mix` module in the header file has been replaced by the declaration of function prototypes. The two functions NCO() and COMP_MIXING() are just copied from their member function equivalents in the SystemC implementation; however, now they have to be declared as non-member functions (also called free functions). The module nco_mix is replaced by a function NCO_MIX(), which will be called by the framework.

```
#include "ofdm_demod_pkg.h"
#include <math.h>
int NCO_MIX(T_FFT* data_r, T_FFT* data_i,
            T_NCOINC* in_nco_inc);
int NCO(const T_NCOINC* incr,
            T_FFT* sin_val, T_FFT* cos_val);
int COMP_MIXING(T_FFT* data_r, T_FFT* data_i,
            T_FFT* sin_val, T_FFT* cos_val);
```

The code snippet below shows the implementation of the NCO_MIX() function. The arguments passed to the function are a pointer to the buffer for the real input data, a pointer to the buffer for the imaginary input data and a pointer for the increment of the NCO, which is determined by the fine_freq module. There are no explicit pointers to the output buffers used by the NCO_MIX function. This is because the processing is done in place, in other words the samples in the input buffer are overwritten by the output samples. The same applies to the COMP_MIXING function, which used to have two additional arguments for the output data in the high level model (see page 74).

The only task handled by the NCO_MIX() function in the example is to call the two functions that were previously called by the process main() of the nco_mix module. Access to the sc_fifo<> ports of the module has also been removed from the original process code, because communication between the modules is handled by the framework in the ANSI C implementation.

```
#include "nco_mix.h"
int NCO_MIX(T_FFT* data_r, T_FFT* data_i,
            T_NCOINC* in_nco_inc)
{
  /*Internal variables*/
  T_FFT nco_cos;
  T_FFT nco_sin;
  NCO(in_nco_inc,
      &nco_cos, &nco_sin);
  COMP_MIXING(data_r, data_i,
              &nco_cos, &nco_sin);
  return(1);
} /*End of the function definition*/
```

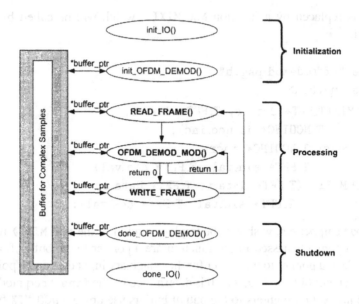

Figure 3.7. Structure of the framework used by the ANSI C implementation of the OFDM demodulator

3.4.3 Replacing the FIFO Channels

The structure of the framework used to replace FIFO channels and the SystemC scheduler is shown in figure 3.7. As can be seen from the diagram, there are three different phases of execution, *Initialization*, *Processing*, and *Shutdown*. All three phases use a pointer to access the same buffer, which is used to store complex samples for processing by the demodulator. The buffer_ptr shown in figure 3.7 is a pointer to a structure that comprises a pointer to the buffer for real samples, a pointer to the buffer for imaginary samples, a counter for the number of processed data blocks, and a field for the number of samples that can be stored in the buffer. This allows for an easy adoption of the buffer size for implementation on different architectures.

During initialization the function init_IO() handles the setup of the I/O devices. In the current implementation the I/O devices are simply two files, one for reading input data and one for writing output data. Other I/O devices could be added by re-implementing the init_IO() function and other device dependent functions used by READ_FRAME() and WRITE_FRAME(). The init_OFDM_DEMOD() function is used to dynamically create storage space for the buffer and to initialize global variables of the demodulator.

The first action in the processing phase is to read a frame from the input device. The READ_FRAME() function does not directly access the underlying input device. To achieve maximum flexibility a layered approach is used that separates the application specific functions and the device specific functions. READ_FRAME() only calls a generic function to acquire data from the input device, which has to be re-implemented for the given device.

After having read a frame the processing takes place. The FSM, used to schedule the function calls for the different blocks of the demodulator, is implemented by OFDM_DEMOD_MOD(). The function returns control to the framework after each algorithmic block and is consecutively called by the framework until all blocks have been run on the data stored in the buffer.

The WRITE_FRAME() function, as the last stage of the processing phase, writes the processed samples to the output device. It has the same layered structure as already discussed for the function READ_FRAME().

The framework continues with the shutdown phase after all frames available on the input device have been processed. The memory that was dynamically allocated by init_OFDM_DEMOD() is now de-allocated by done_OFDM_DEMOD(). The I/O devices are finally closed using the function done_IO().

3.4.4 Performance Comparison

In this section, the results of executing the ANSI C description of the OFDM demodulator on the A4 processor are presented along with a comparison of the results obtained from operator counting (see section 3.3.3).

The first thing to note when comparing operator count with real run time is that the results from counting operators will always be inaccurate up to a certain degree. That is the case because only the processing power required for the arithmetic operations in the data path is measured, whereas the operations needed for control structures like branches or loops are ignored. Another effect that is not completely taken into account is the overhead introduced by pipeline stalls and cache misses.

To yield good accuracy, despite the factors discussed above, the run time for an arithmetic operation on the A4 was measured in a context that is as close as possible to the real program. The accumulated run time for every operator was measured inside a 'for loop' including access to array elements (see code snippet below). As the processing inside the OFDM demodulator is block based, dividing the accumulated run time by the number of loop iterations should deliver reasonable estimates of the run time for an operation in the context of the demodulator.

```
void mul(T_DATA* x_ptr, T_DATA* y_ptr) {
  extern T_DATA z[L];
  int i;
  for(i = 0;i < L;i++)
    z[i] = *(x_ptr+i) * *(y_ptr+i);
}
```

Operation	run time [clock cycles]
+	166
*	755
sin	9,000

Table 3.5. Run time of various operations on the A4 processor. All operations use double precision floating point emulation. Run time is measured in clock cycles required by the processor and includes the overhead of array access to retrieve the operands.

Using this technique, run time estimates for various operations on the A4 have been obtained. Table 3.5 shows the run times for some operations for the data type double. As the A4 processor used in this project does not have a floating point unit, all operations were carried out using floating point emulation. Therefore the run time required by those operations is much higher than the run time of the corresponding fixed point or integer operations. Finally, it should be noted that the run time required for one floating point operation depends on the input data. The cycle count presented in this article for the various operations has been obtained using data that results in a high run time, and therefore presents a worst case estimate.

The first example that will be studied is the low pass filter. A histogram obtained from counting the operations in the low pass filter has already been shown in figure 3.5 for processing a complete frame. To make things simpler the comparison has been carried out using only 2,872 samples instead of a profile for a complete frame. The reason for choosing exactly 2,872 samples is based on the internal implementation details of the OFDM demodulator architecture and lies beyond the scope of this article.

Operation	# of calls	cycles/operation	# of cycles
+	74,672	166	12,395,552
+=	80,416	156	12,544,896
*	80,416	755	60,714,080
total cycles			85,654,528
measured cycles A4			53,300,060

Table 3.6. Straightforward calculation of run time estimates from operator count for the low pass filter

Table 3.6 shows a straightforward calculation of the run time estimate for the low pass filter by simply multiplying the number of operations by the estimated run time. It is immediately evident that this seems to deliver a useless result, because the estimate exceeds the real run time by nearly 60 percent.

Operation	# of calls	cycles/operation	#of cycles
+	40, 208	166	6, 674, 528
+0	34, 464	14	482, 496
+=	45, 952	156	7, 168, 512
+=0	34, 464	14	482, 496
*	45, 952	755	34, 693, 760
*0	34, 464	14	482, 496
total cycles			49, 984, 288
measured cycles A4			53, 300, 060

Table 3.7. Run time estimate calculation for the low pass filter taking into account the implementation of the filter

The result can, however, be much improved by considering the implementation of the low pass filter. The filter is implemented as a half band filter in which half of the coefficients are zero. Therefore half of the operations result in a multiply or add with a zero argument. Performance measurements for multiply and add operations with zero argument show that the time required for those operations is much less than the time that is used for an operation with non-zero argument. Table 3.7 shows the results taking into account the shorter run time for operations with zero argument. Now the estimate is only 6 percent apart from the real run time.

data type	cycles for multiplication	cycles for addition	deviation of estimate from run time
double	755	166	-6%
float	275	136	-15%

Table 3.8. Degradation of estimation results for shorter operator run times

Of course, a different implementation of the half band filter could have been chosen for the high level model where the operations with zero arguments are avoided. However, in this project a generic implementation of a Finite Impulse Response (FIR) filter was used in the system-level model and directly re-used in the ANSI C model.

The discrepancy of straightforward calculation and measurement for the simple FIR filter example illustrates the limitation of the predictions that can be made with the operator counting technique. If the run time of an operation depends on the input data then it is generally difficult to make an exact run time prediction without knowing the implementation and also the distribution of the expected input data. Another limitation is that to deliver precise results, the run time for arithmetic operators must be much bigger than the run time required for other instructions in the program such as loops or branches. This becomes obvious from having a look at table 3.8, which shows the degradation of the run time estimates for shorter operator run times.

Operator counting in combination with run time estimates for the individual operations can be used as a tool to predict the run time of the system on the target platform. However, it has to be used with care and with the knowledge of its limitations. Another way to exploit the knowledge of the number and types of operations per module is to calculate a measure for the numerical complexity of the individual modules. This can be used to identify the modules with the highest processing load, which will then be candidates for either further optimization of the implementation or for implementation in hardware.

module name	relative numerical complexity
fine_freq	1
nco_mix	11.3
lp_filter	6.6
sym_frame_syn	2.1
fft	2.4
post_proc	5.3

Table 3.9. Relative numerical complexity using the fine frequency module as a reference

The calculation of the overall numerical complexity of a module starts with identifying the operator that imposes the least computational burden (in this case operator+), and calculating relative run times for all the other operations. The number of each type of operation in a module is then scaled by the appropriate relative run time, and the results are accumulated to the total numerical complexity (relative to the operator+). The results are shown in table 3.9. It can be seen that the sample based blocks impose the largest computational burden.

Because of the large computational power required by the low pass filter and the complex mixer, these two modules are candidates for being implemented in hardware. In the following it is assumed that the A4 processor only handles the block based processing comprised by sym_frame_syn, fine_freq, fft and post_proc. Measurements using an A4 evaluation platform have shown that a theoretical clock frequency of about 70 GHz would be required to run the block based processing in real time. This is far from being feasible.

Results from measuring the run time for operations for various data types have shown that a speed up by a factor of 15 is realistic for a conversion from double precision floating point to fixed point. That would result in about 4–5 GHz necessary clock speed. Experience from previous projects shows that this could be reduced by a further factor of 2–4 if hand optimized code with assembly language for the critical parts is used. To further reduce the required clock speed, more modules could be implemented in hardware, for example the post_proc module. That approach will not be further discussed in this paper.

3.5   Further Refinement — Operator Grouping

At this point in the refinement process the type of required operators as well as their usage statistics is known from operator counting (see 3.3.3 and 3.4.4) in the SystemC model. In addition a realistic estimate for the performance exists from the ANSI C port to a target processor. The next step, as discussed throughout this section, is using SystemC to derive and assess hardware add-ons to the baseline processor to lower the number of processing cycles required for the OFDM demodulator.

3.5.1   Introduction

As mentioned earlier, the target processor is the ARCtangent™-A4 [ARC Int., 2001]. This customizable 32 bit processor core is a soft macro and can be implemented on virtually all ASIC and FPGA technologies. The user may select — under the control of a vendor provided configuration tool — different hardware options for the processor core and generate a human-readable VHDL or Verilog source database that includes all modules of the core. Configurable options include the system interfaces, size and organization of instruction and data caches as well as details of the instruction set.

The performance figures in table 3.6 to 3.8 are derived from measurements on the baseline processor. In addition, hardware extensions for a barrel shifter, normalize, and 32 bit multiplier may be easily added. These extensions are known to the compiler engine and suitable instructions are inferred when the code is translated. Except address calculation that may be carried out in parallel, the parallelism in core is limited to one operation at a time. Using the mentioned extensions and assuming fixed point data types and an optimal scheduling done by the compiler, each simple operation (add, sub, multiply, assignment) executes in one cycle.

3.5.2   HW/SW Partitioning

Following a traditional ASIC design flow, the next step in the refinement process would be a module by module transition to a synthesizable behavioral or RT level representation. In this context SystemC is used similar to a conventional hardware description language like VHDL or Verilog. An introduction on using SystemC for hardware implementation can be found in [Synopsys, 2002c] or [Bhasker, 2002].

Applying this method to all modules of the OFDM model and using an appropriate synthesis tool, the result is a 'hardware only' implementation of the OFDM demodulator. Such an implementation is presented in [Nagel et al., 2001], using VHDL and Synopsys Behavioral Compiler for synthesis. Besides some tool and language specific details, the general flow for SystemC is expected to be similar to the one presented in [Nagel et al., 2001]. Also experience shows that the results from SystemC based RTL or behavioral synthesis are approximately the same as for traditional HDLs (see for example [Niemann and Speitel, 2001] for a comparison of SystemC and VHDL for RTL design). This flow will not be discussed in detail in this article.

Based on the characteristics of the OFDM demodulator and the results from the previous chapters, a suitable HW/SW architecture [Sigwarth et al., 2002] is shown in figure 3.8. Some of the blocks of the data flow model could be transferred into dedicated hardware (type A extensions) or co-processor IPs (type B extensions). Others would be preserved as software that is executed on the processor. For example, the sample based nco_mix and lp_filter are candidates for type A extensions shown in 3.8. The SystemC code is refined to RT-level SystemC, and the FIFO based interfaces and communication are replaced by some handshaking protocol or an interface to the software running on the processor.

The block based fft and post_proc modules might be implemented as loosely coupled co-processor IPs (type B extensions in figure 3.8), whilst the closure of the fine_freq control loop and the sym_frame_syn synchronization would be done by software.

The approach taken in this article targets for a more balanced HW/SW partitioning with finer granularity. Operations that require too much run time are implemented as specialized instruction set extensions for the given processor core. As shown in figure 3.8, such instruction set extensions (type C) will ex-

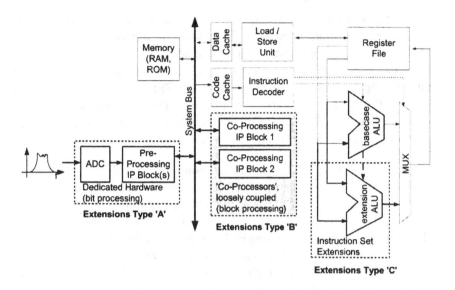

Figure 3.8. HW/SW architecture of the OFDM demodulator

ist in parallel to the original processor ALU and implement the logic for the additional operator(s). Data paths and control logic are shared with the core processor; execution is under the control of the ordinary program code.

In this approach to HW/SW partitioning, SystemC is used:

- for the formal description and implementation of the additional operators;

- to re-code the data flow model, making use of the additional operators;

- to re-verify the behavior of the data flow model;

- to evaluate and compare the performance gain of different operator sets;

- finally, to derive a correct by construction ANSI C variant which may be executed on the target hardware and which uses the instruction set extensions.

SystemC offers very fast turnaround times because only minor modifications to the original data flow model are required to implement and test different operator sets.

3.5.3 Custom Operator Example: FFT

System Engineering. The application of user defined operations for the processor core is demonstrated using the fft module as an example. This FFT does a block based processing of input data and consists of radix-2, radix-3 and radix-4 stages. In the original data flow model the data type for the FFT input and output samples is given by the type T_FFT. For system engineering the type T_FFT usually translates to double or float. Given the limited bit width of the data path (32 bits) and the complexity of floating point arithmetic, it is generally not a good idea to keep floating point data types for the target system.

Fixed point Model. To derive a fixed point model a block floating point approach is used: All inputs are scaled by a constant factor s_n to the range $[-1, 1[$ and treated as numbers consisting only of fractions. 15 bit precision is used whilst the 16th bit represents the sign; the real and imaginary parts are packed into one 32 bit word. During FFT processing the intermediate results are statically re-scaled and rounded to avoid overflows and to preserve reasonable resolution. After processing, the correct floating point values are recovered by inverse scaling with the input scale factor s_n and the product of all per-stage scale factors; using standard floating point representation this is done by adding a constant value to the original exponents. For reasons of simplicity, in this article the well known radix-2 FFT operation is presented. An ANSI C code snippet for the butterfly operation is shown below.

```
/* FFT butterfly in ANSI-C */
/* inputs:
   xr (real input vector), xi (imaginary input vector)
   wr (twiddle factor 1), wi (twiddle factor 2)
   outputs:
   yr (real output vector), yi (imaginary output vector) */
tr = wr * xr[i1] - wi * xi[i1];
ti = wr * xi[i1] + wi * xr[i1];
yr[i1] = xr[i0] - tr;
yi[i1] = xi[i0] - ti;
yr[i0] = xr[i0] + tr;
yi[i0] = xi[i0] + ti;
```

3.5.4 Operator Grouping

The next step is to find common structures in the FFT and group several simple operators like add and mul into one or more complex operators. Examples for possible operator groups are given in figure 3.9.

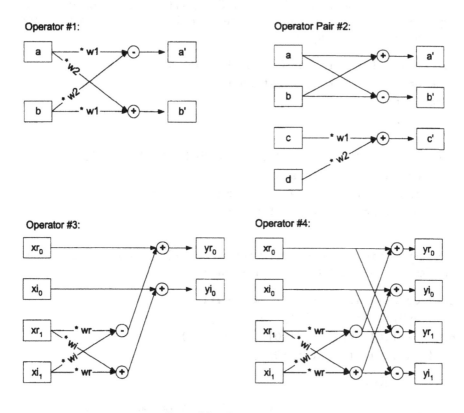

Figure 3.9. Operator groups

	area [gate equiv.]	estd. delay (10ns constraint)
op1	8.201 k	9.63 ns
op21	0.205 k	5.43 ns
op22	4.460 k	9.52 ns
op3	10.487 k	9.88 ns
op4	11.277 k	9.86 ns

Table 3.10. Area estimates for the different operator groups

The examples are sorted by increasing complexity. Example op4 represents a complete complex butterfly similar to the C code snippet given before. Table 3.10 shows the approximate hardware size of each group when constrained to a constant delay of 10 ns (nominal) in a 0.18 um technology.

Having later translation back to ANSI C in mind, the custom operators are implemented as ordinary functions. The functions for operators op21 and op22 are shown in the code snippet below:

```
void op21 (T_FFT a, T_FFT b, T_FFT *ao, T_FFT *bo)
{ /* operator pair 2, first op */
  *ao = a + b;
  *bo = a - b;
}
T_FFT op22 (T_FFT c, T_FFT d, T_TWIDDLE w1, T_TWIDDLE w2)
{ /* operator pair 2, second op */
  /* fmul emulates fixed point multiplication */
  return ( fmul(c, w1) + fmul (d, w2) );
}
```

Using the two custom operator functions op21() and op22(), the original FFT butterfly code may be rewritten as follows:

```
tr = op22(xr[i1]), xi[i1], wr, -wi);
ti = op22(xi[i1], xr[i1], wr, wi);
op21 (xr[i0], tr, &(yr[i0]), &(yr[i1]));
op21 (xi[i0], ti, &(yi[i0], &(yi[i1]);
```

3.5.5 Results

A setup identical to the one in chapter 3 is used for operator counting. Table 3.11 gives the results for different combinations of the 4 operator groups defined in figure 3.9.

	case A	case B	case C	case D
op1	2, 304			
op21	4, 608	4, 608		
op22		4, 608		
op3			4, 608	
op4				2, 304
total op. count	6, 912	9, 216	4, 608	2, 304
total area [gate equiv.]	8.4 k	4.7 k	10.5 k	11.3 k

Table 3.11.   Operator count vs. area

As can be seen from the table, case B has the highest operator (and thus clock cycle) count but the smallest area requirements. Case D provides the best speed up, although this also corresponds to the highest area. Depending on the cost function — trade off between speed and area — that is applied, case B may be of interest, whilst case C doubles, compared to case D, the processing time by providing only a 10% area gain.

Another important fact that affects the selection of optimal operators is the number and width of available data paths. In the target processor 2 independent 32 bit data paths for 2 input arguments are routed to the ALU, while only one 32 bit bus exists for the result. Using 16 bit precision and considering all arguments for the different operators as variable, actually only op1, op21 and op22 may be implemented, while op3 or op4 would require 6x16 bit (xr0, xi0, xr1, xi1, wr, wi) and thus exceed the capacity of the input data paths. Moreover, op4 produces results of 4x16 bits (yr0, yi0, yr1, yi1), thus exceeding the capacity of the data bus used for results.

If op3 and op4 are only used for FFT processing, wi and wr may be assumed to be semi-static. In fact, wi and wr are constants (twiddle factors), usually taken from pre-calculated sine and cosine tables. They vary from butterfly to butterfly, but are independent of the data that is processed by the FFT. In other words, wi and wr may be generated internally in op3 or op4, for example by table lookup, and there is no need to explicitly provide wi and wr as arguments. This limits the input width to a total of 64 bits and enables the implementation of op3 in the given target hardware. The implementation of op4 would require additional modifications of the processor core that are beyond the scope of this article.

3.6   Summary

Modeling and refinement using SystemC was demonstrated in the preceding sections by means of an OFDM demodulator example. Here the techniques used to take advantage of the modeling capabilities offered by SystemC are briefly summarized.

The high level model is used to explore the performance of various algorithms in terms of bit error rate as a function of signal to noise ratio. Furthermore, it serves as a basis for further refinement and as golden reference model for all subsequent steps. It must be possible to simulate a large amount of data within reasonable time in order to get a reliable measure for the performance. Moreover, algorithms should be easily interchangeable to allow for seamless algorithmic exploration. Finally, it should be possible to get to lower levels of abstractions as easy as possible.

To achieve the abovementioned objectives we have applied the following modeling principles:

- *Functional decomposition.*   Place every functional block (algorithm) inside a single SC_MODULE. This enables easy replacement of single functional blocks by alternative implementations;

- *Minimal interfaces.* Minimize the number of ports of the module. Put the synchronization inside the communication channels, for instance use sc_fifo<> channels instead of hardware signals for data and handshaking. This makes changes of the model structure during development more manageable;

- *Identify and name different data types.* Use typedefs as much as possible. Even in the high level model, where double will be the predominant data type, try to identify different categories of data that might be implemented with different bit widths later on. This greatly simplifies the transition from a floating point to a fixed point model;

- *Read, process, write.* Inside a process the first step should be to read all input ports to temporary variables. Based on these temporaries do the processing (perform the algorithm). Finally, write back the data to the output ports. On the one hand this makes integration of algorithms written in pure C into a SystemC module particularly easy. On the other hand, it allows a SystemC model to be easily reverted to a pure software implementation;

- *One single process per module.* Every SC_MODULE should have a single SC_THREAD process. This is no strict rule, but in general simplifies the transition from the data flow model to a implementation model, as there is no synchronization between parallel processes that has to be taken into account during refinement;

- *Place algorithm inside a function.* It is a good idea not to directly 'inline' the algorithm code inside the SC_THREAD process. Instead, it should be placed inside a member function of the module and be called from the SC_THREAD process. This simplifies subsequent refinement;

- *Use block based interfaces for modules with block processing.* There are two benefits from this rule. Firstly, the number of calls to read() and write() is reduced. Secondly, a change of the block size is possible without having to modify the module itself;

- *Minimal synchronization effort.* Try to minimize calls to the simulation kernel. This is achieved by using sc_fifo<> channels instead of sc_signal<> channels and by choosing appropriate FIFO sizes. The less the synchronization effort, the faster the simulation;

- *Do refinement in small steps.* To exploit the unified environment offered by SystemC do not try to achieve everything in one giant step. With the OFDM demodulator, for example, we have used the ANSI C description as an intermediate step;

- *Separate communication refinement from behavior refinement.* Derived from the principle *Do refinement in small steps.* In our example the first step was to develop the framework (communication refinement) and the second step to modify the FFT implementation (behavior refinement);

- *Verify every step.* Use the test bench of the high level model to verify every refinement step against the golden reference. This often requires the additional implementation of adapters to connect incompatible interfaces. The adapter implementation should be considered as integral part of the refinement step. Obey the principle of *Minimal interfaces* to make adapter implementation more easy.

3.7 Conclusions

In this article we have presented refinement of a high level SystemC model towards embedded software, dedicated hardware, and CPU instruction set extensions.

Most individual steps we presented could have been carried out with conventional hardware and software design methods using traditional languages like C, C++ and VHDL. But, with SystemC, a single language, environment, and methodology may be applied throughout the complete refinement process. In addition, SystemC offers some unique features that enable a smooth transition from higher levels of abstraction to the lower ones.

In the high level model we have made use of the encapsulation in modules, the connection by FIFO channels and the synchronization and scheduling offered by the built in simulation kernel. SystemC, moreover, helped in retrieving the number and type of operations for each module of the high level model. In contrast to commercial data flow modeling systems using proprietary languages, the SystemC model can be run on a large variety of platforms without having the need of installing additional software.

During the transition to ANSI C we had the advantage of a unified development environment. Moreover, the task could be split in small manageable steps that could always be verified against the high level model.

New complex operators were easily implemented and verified using SystemC and the high level model. The performance gain of these operations could also be evaluated using SystemC. The last step would have been to implement those custom operators in hardware using SystemC for RT level description. This step was not carried out because the feasibility of hardware implementations using SystemC has already been proven (see, for example, [Niemann and Speitel, 2001]).

Chapter 4

An ASM Based SystemC Simulation Semantics

Wolfgang Müller[1], Jürgen Ruf[2], Wolfgang Rosenstiel[2]

[1] *Paderborn University, C-LAB*

[2] *Tübingen University, Department of Computer Engineering*

Abstract   We present a formal definition of the event based SystemC V2.0 simulation semantics by means of distributed Abstract State Machines (ASMs). Our definition
provides a rigorous and concise, but yet readable, definition of the SystemC specific operations and their interaction with the simulation scheduler that covers
channel updates, notify, notify_delayed, wait, and next_trigger operations. We
present the semantics in the form of rules by means of distributed ASMs reflecting
the lines of the SystemC V2.0 Standard Manuals and reference implementation.
The semantics introduced is defined to complement the language reference manual with a precise definition reflecting an abstract model of the SystemC reference
implementation, which can be used for advanced applications and for investigat-
ing interoperabilities with other languages.

4.1   Introduction

SystemC is the emerging *de facto* standard for system level modeling and
design from the Open SystemC Initiative (OSCI). In 2002, SystemC received a
major revision and upgrade to Version 2.0 which provides the stable basis for future versions. With this upgrade the main principles were clarified by a generalization, e.g., of the underlying event based synchronizations and channel based
communication concepts. SystemC V2.0 currently comes with well-written
manuals [OSCI, 2002a, OSCI, 2002b], a stable reference implementation, and
complementary text books, e.g., [Grötker et al., 2002]. However, the precise
meaning of several parts of the underlying concepts cannot be easily captured
since natural language descriptions often lack precision. Thus a precise semantics is mandatory for advanced SystemC application in simulation, synthesis,
and formal verification.

W. Müller et al. (eds.),
SystemC: Methodologies and Applications, 97–126.

In this chapter, we present a concise and rigorous but yet intuitive semantic definition of SystemC Version 2.0 by Gurevich's *distributed Abstract State Machines* [Gurevich, 1995]. ASMs allow us to define the semantics following the lines of the SystemC manuals and the reference implementation. The definition given herein is a significant major revision of the semantics given in [Müller et al., 2001] towards the event-based interaction of the simulation scheduler and the user defined processes (methods and threads). We develop a mathematical definition of SystemC in terms of a *SystemC Algebra* with the precise semantics of channel updates as well as wait(), notify(), notify_delayed(), and next_trigger() operations.

We mainly introduce our formal semantics as a precise and concise but yet intuitive definition which complements the SystemC language reference manual and the reference implementation. Its main application is as a basis for the language studies and interoperabilities with other languages as well as for reasoning about SystemC models, i.e., for formal verification and synthesis. Studies in language interoperabilities are supported by our ASM definition since comparable definitions are available for VHDL, Verilog, and SystemC as it is outlined in the next section.

The remainder of this chapter is organized as follows. Section 2 discusses related works. In Section 3, we briefly introduce what is needed from distributed ASMs without going into theoretical details. Section 4 gives an overview over the general principles of SystemC before Section 5 and 6 introduces the semantics of the individual SystemC specific operations as well as the SystemC scheduler. Section 7 gives an example that executes on the introduced Abstract State Machines before the conclusion closes the chapter.

4.2   Related Works

Over recent years research in the formal semantics of hardware description languages has been dominated by VHDL. There were numerous approaches based on temporal logics, functional semantics, denotational semantics, and operational semantics applying Boyer–Moore Logic, Process Algebras, Petri Nets, and many other means [Delgado Kloos and Breuer, 1995]. Most of the approaches cover VHDL subsets dedicated to application in formal verification. In 1995 the simulation semantics of VHDL'93 was also introduced by Abstract State Machines in [Börger et al., 1995] as a complementary precise, formal, but yet readable, documentation to capture the relevant principles of non-trivial interaction of the user defined VHDL processes and the VHDL simulation kernel. That definition was extended towards VHDL-AMS in [Sasaki et al., 1997] and used for the investigation of VHDL/Verilog interoperabilities in [Sasaki, 1999]. Later, the ASM approach was applied to define the complete execution semantics of SpecC in [Müller et al., 2002]. In addition ASMs have been successfully applied to the definition of semantics of various programming languages, like Java [Börger and Schulte, 1998] and C++[Wallace, 1995]. The ITU standard SDL 2000 was the first standard where the complete dynamic semantics is described by means of ASMs [Glässer et al., 1999].

Considering the formal semantics of SystemC, there are only very few activities today. We have applied the ASM approach to define the formal semantics of behavioral SystemC V1.0 [Müller et al., 2001] that was defined through similar patterns as the VHDL'93 semantics given in [Börger et al., 1995]. More recently the denotational semantics of synchronous SystemC V1.0 was introduced for application in formal verification [Salem and Shams, 2003].

In this article we define a replacement of our formal SystemC V1.0 semantics introduced in [Müller et al., 2001]. When advancing from V1.0 to V2.0 significant changes and generalizations were introduced in SystemC V2.0, such as a complete redefinition of the simulation scheduler and additional statements like immediate notification. SystemC semantics has advanced from a control-oriented description in [OSCI, 2000] to a complete event based definition in [OSCI, 2002a, Grötker et al., 2002], which also required an essential update of our previously introduced ASM definition in [Müller et al., 2001]. Owing to the complete event based principles of the language, the essential concepts of the quite complex interaction of the scheduler and the user defined processes are not easy to capture.

4.3   Abstract State Machines

Gurevich initially introduced basic Abstract State Machines (ASMs[1]) in 1991 [Gurevich, 1991]. A revised definition of ASMs with various extensions, commonly known as the Lipari Guide, was published in [Gurevich, 1995]. Whereas Gurevich has originally defined ASMs for considerations in complexity theory, multiple publications have demonstrated the applicability of ASMs for formal specification for various purposes. Examples come from hardware architectures, software architectures, communication protocols, and programming languages [Börger and Stärk, 2003].

An ASM specification is a program executed on an abstract machine. The program comes in the form of guarded function updates (rules). Rules are nested if–then–else clauses with a set of function updates in their body. Based on these rules, an abstract machine performs state transitions with algebras as states. Firing a set of rules in one step performs a state transition. Only those rules are fired whose guards (*Condition*) evaluate to true. Rules are of the form

if *Condition* **then** $<$ *Updates* $>$ **else** $<$ *Updates* $>$ **endif** .

---

[1] Formerly known as Evolving Algebras

At each step the guards evaluate to a set of function updates (block) each of the form

$$f(t_1, ..., t_r) := t_0$$

where t_i are terms (including functions). A block is a set of function updates separated by a comma [2]. The individual function updates of each block are collected in a so-called update set. The individual updates of the update set are simultaneously executed in one step. That means that all updates in a block are simultaneously executed. Each function update changes a value at a particular location, which is given by the left hand side of the assignment. Functions are considered as global. In the classical ASM definition, two or more simultaneous updates of the same location in one update set define inconsistency and no state transition and no update in the update set is executed. In this chapter, we take a slightly modified ASM definition. In the case of more than one update on the same location we non deterministically choose one and remove the others from the update set.[3]

The following example illustrates a guarded update of a block with two update instructions:

if *Condition* **then** $A := B, B := A$ **endif**

It defines the simultaneous update of the 0-ary functions A and B. Since both updates are simultaneously executed, A becomes the value of B and vice versa. The rule fires when *Condition* evaluates to *true*.

ASMs distinguishes *internal* and *external functions*. An internal function has a well defined signature and mapping. An external function non-deterministically maps to a valid value, i.e., an external function may return different results when given the same arguments. External functions are often used when parts of a system are left unspecified. External functions are typically applied for modeling the system's environment.

ASMs are multi-sorted based on the notion of universes. We assume the standard mathematic universes of Boolean, Integer, List, etc., as well as the standard operations on them without further mentioning. A universe can be dynamically extended with individual objects by

extend *Universe* **with** v $<Rule>$ **endextend** ,

where v is a variable which is bound by the **extend** constructor.

---

[2] Note here that Gurevich [Gurevich, 1995] does not introduce a special symbol for separating updates in a block. We use a comma as an explicit block separator in this chapter.

[3] This modification has no impact on the theory of ASMs rather than helps us to keep the definition of SystemC simulation kernel simpler.

The **choose** constructor defines an arbitrary selection of one element in a universe

$$\textbf{choose } v \textbf{ in } Universe < Rule > \textbf{ endchoose },$$

where v is non-deterministically selected from the given universe. The **choose** constructor can be qualified by a condition (**satisfying**). v is *undef* when the condition evaluates to *false*.

The **var** rule constructor defines the simultaneous instantiation of a rule:

$$\textbf{var } v \textbf{ ranges over } Universe < Rule >$$

Executing the constructor means to spawn and execute the rule for each element in *Universe* simultaneously, i.e., the constructor basically spawns n rules where n is the number of elements in *Universe*. The following example demonstrates the application of this constructor. It defines a rule which specifies that each non-empty l from the universe $LIST$ is replaced by the list's tail, i.e., deleting the first list element. l refers to any valid instance of $LIST$.

$$\textbf{var } l \textbf{ ranges over } LIST$$
$$\textbf{if } l \neq \langle \rangle \textbf{ then } l := tail(l)$$

The definition of our SystemC semantics is defined by distributed ASMs. Distributed ASMs consist of a collection of autonomously operating agents interacting with each other by reading and writing shared locations of global system states. The underlying semantic model regulates such interactions so that potential conflicts are resolved according to the definition of partially ordered runs. Distributed ASMs partition rules into modules where each module is given by its module name ν. A module is instantiated to execute by setting $Mod(a) := \nu$ for an agent a. The symbol *Self* refers to a after the instantiation.

4.4 SystemC Basic Concepts

The SystemC language is a C++ language superset. SystemC comes as a C++ library with a header file defining classes and a link library with a simulation scheduler. Any ANSI C++-compliant compiler can compile SystemC. In the current SystemC V2.0.1 reference implementation the additional constructs are defined as macros, C++-classes, and C++-methods. When compiling a SystemC program the resulting executable first instantiates the SystemC modules and then starts a cycle based simulator that allows high speed simulation with integrated simulation control facilities. SystemC has received a significant revision from V1.0 to V2.0. Since no major additional changes in the basic principles of SystemC V2.0 with respect to future versions are expected, we can focus our investigations on SystemC 2.0 without loss of generality. In this section, we briefly introduce the structural aspects of SystemC and outline the basic behavioral aspects of the language.

4.4.1 Structural SystemC Descriptions

In SystemC the fundamental building block is a module (SC_MODULE). Modules may contain modules or functional units, i.e., processes, which are distinguished into SystemC methods and threads. Modules are connected via ports, interfaces, and channels. Channels are distinguished in primitive channels like Signals, Fifos, etc., and hierarchical channels to implement complex communication mechanisms like 'on chip buses'. Ports are objects connecting channels and modules. Though they are separate individual objects, they can be viewed as pointers to channels.

In a SystemC description sc_main denotes the main module which recursively instantiates a set of hierarchically embedded submodules. The following code gives a very small example, with $p1$ as an instance of module $m1$ and $p2$ as an instance of $m2$. Signal i connects the two ports X and Y of the instances before sc_start invokes the simulator for at most 20 time units.

```
int sc_main(int argc, char** argv) {
  sc_signal<bool> i;
  m1 *p1 = new m1("m1");
  m2 *p2 = new m2("m2");

  p1->X(i);
  p2->Y(i);
  sc_start(20);
}
```

The hierarchical instantiation of modules from the beginning of sc_main to the call of the simulator is denoted as the *elaboration phase* of the SystemC execution.

4.4.2 SystemC Simulation

After elaboration, derived from hierarchically organized modules, SystemC establishes a network of a finite constant number of parallel communicating processes $p \in PROCESS = METHOD \cup THREAD$[4], which under the control of the distinguished simulation scheduler updates *CHANNELs* and *VARIABLEs*. As other distinguished objects, *CLOCKs* are introduced as a combination of *SIGNALs* and a *MODULEs*. *CLOCKs* can be determined as predefined thread with a unique periodically changing value. Owing to the

---

[4] SystemC actually introduces threads and cthreads (clocked threads) where the latter are mainly introduced for the matter of efficient simulation and synthesis. Since SystemC V2.0 unifies the behavioral semantics of threads and cthreads, cthreads become a specialization of threads, so that we can consider the semantics of threads here without the loss of generality.

concepts of delayed assignments, channels do not change their values immediately. Writing to a channel schedules an update request according to which
the simulation scheduler performs an update() on the channel later so that a
write operation does not become effective before.

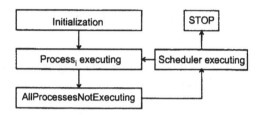

Figure 4.1. SystemC process scheduler interaction

The simulation scheduler is invoked by the sc_start() command and performs an initialization which sets the current simulation time T_c to 0 and executes all processes until they suspend except those for which the SystemC
function dont_initialze() is called. After initialization there is a mutually
exclusive execution of the simulation scheduler and the pseudo-concurrently
running user defined processes as given in Fig. 4.1. Pseudo-concurrent means
that from the set of *ready* processes one after the other is selected and executed,
where the order of selection is deterministic but implementation dependent. The
scheduler periodically starts its execution as soon as all user defined processes
are not *executing*, and incrementally selects and executes a process which is
ready to run until all processes are *suspended*, i.e., there are no more processes
which are *ready* to run. A SystemC thread is *executing* and evaluating its sequential statements until reaching a wait() operation, upon which it changes to
suspended. An *executing* SystemC method becomes *suspended* each time after
executing the last statement. For suspension and invocation, SystemC distinguishes static and dynamic sensitivity based on EVENTs, where the static sensitivity is the default mode. When suspending, the static sensitivity list, which
is defined at elaboration time, is taken as the default sensitivity list. Dynamic
sensitivity is assumed when the static sensitivity list is explicitly overwritten by
a local dynamic sensitivity list, which can be given by the definition of an event
expression, a timeout, or a combination of both. Owing to the different states
of the process life cycle, we set *status(p)* ∈ *{executing, suspended, ready}* for
each process p (cf. Fig.4.2).

This defines the basic life cycle of a SystemC process as given in SystemC
V2.0 so far. SystemC additionally covers exception handling by the means
of watching statements. Fig. 4.3 sketches the extension of the basic life cycle
towards watching conditions. The focus of this chapter is on the basic life cycle.

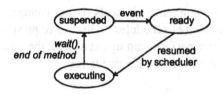

Figure 4.2. Life cycle of a process

For the enhanced ASM semantics including watching conditions, the reader is referred to [Müller et al., 2001].

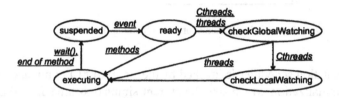

Figure 4.3. Life cycle of a process with exception handling

In SystemC V2.0 the interaction between the scheduler and the user defined processes is completely based on EVENTs. Processes have explicit notification (i.e., generation of events) by *notify* and *notify_delayed*. Suspensions on time-outs and channel updates generate internal events. The list of pending events determines the trigger upon which processes become *ready* with respect to the current simulation time T_c. This defines an underlying discrete SystemC time model in which EVENTs are ordered by their time components with respect to T_c.

The individual tasks of a scheduler are given as phases which are divided into steps. After initialization the simulation scheduler continuously iterates through two different phases: evaluate and update (cf. Fig. 4.4). The evaluate phase defines the invocation of processes and evaluation of their statements within one delta cycle (see (1) in Fig. 4.4/5). When no more processes are *ready*, the scheduler advances to the next delta cycle and proceeds to the update phase. In update, it decides either on the next delta cycle without time advancement (see (2) in Fig. 4.4/5), or to proceed to the next time cycle (see (3) in Fig. 4.4/5). The decision is based on the execution of different steps as given in Fig. 4.5, where the Step 2–3 correspond to the evaluate phase and Step 4–8 to the update phase. In update, the scheduler first checks for *update_requests* on CHANNELs which were scheduled by processes when future value changes on CHANNELs are requested. For each *update_request* the corresponding *update()* is executed on a specific CHANNEL where *update()* can be individually overloaded for each

CHANNEL type. For signals the predefined *update()* assigns the new value to the current one and schedules an update event to the pending events. Here, not a new event of the signal is generated, rather that the already existing default event, which is predefined for each signal, is scheduled. After the update of channels the scheduler checks for events at current time T_c and returns to the evaluate phase when an event is detected. Otherwise T_c is advanced, which activates events for the new time before returning to the evaluate phase.

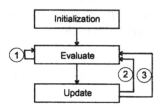

Figure 4.4. Phases of the SystemC simulation scheduler

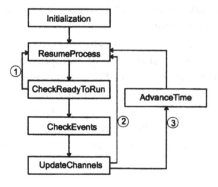

Figure 4.5. Steps of the SystemC simulation scheduler

For our ASM definition, the rules in the next two sections constitute two ASM modules. In the next section, we first define rules for the *PROCESS_ASM-module*, which give the semantics of distinguished SystemC operations by the ASM rules with labels P1–P9. Thereafter the *SCHEDULER_ASMmodule* is defined by the rules labelled S1–S10. When instantiating a program we create one agent instance of the latter for the simulation scheduler and one ASM agent instance of *PROCESS_ASMmodule* for each user defined process.

4.5   SystemC Operations

When evaluating statements of SystemC processes we introduce the notion of a program counter. The ASM function *programCounter* of each process initially points to the first statement of each process. After having processed the last statement, a *THREAD* terminates by setting *Mod(Self)* to *undef*, which terminates the corresponding ASM agent. Methods change to *suspended* after executing the last statement and reset the counter to their first statement. In order to express that a user defined process *Self* executes a statement only when it is in *status executing* and the *programCounter* is assigned to the specific statement, we use the following abbreviation.

$$Self\ executes\ \langle \texttt{operation} \rangle \equiv$$
$$programCounter(Self) = \langle \texttt{operation} \rangle \wedge status(Self) = executing$$

4.5.1   Primitive Channels

As already introduced, SystemC distinguishes primitive and hierarchical channels. When writing to a primitive channel c by $\texttt{c.write()}$ its value is not immediately updated. Its update is based on a delayed update concept, where the process schedules an update request and the simulation scheduler performs the actual update later.

For each primitive channel type such as fifo or signal, $\texttt{write}$ performs differently. Since signals are the most frequently used primitive channels, we focus on the semantics of signal updates here. For signals $\texttt{write()}$ executes the function *request_update()* which schedules an update request for signal c. Further on we assign the value of the given expression *val(Expr)* to the *new_value* of the signal c and finally advance the program counter. Note here that (pseudo-) parallel write accesses to the *new_value* may appear which are resolved by 'last one wins'.

$$
\begin{array}{l}
\textbf{P1}: \\
\textbf{if}\ Self\ executes\ \langle \underline{\texttt{c.write}}(\texttt{Expr}) \rangle \\
\textbf{then}\quad \textbf{if}\ val(Expr) \neq value(c)\ \textbf{then}\ request\_update(c)\ \textbf{endif}, \\
\qquad\quad new\_value(c) := val(Expr), \\
\qquad\quad programCounter(Self) := nextStmt(Self) \\
\textbf{endif}
\end{array}
$$

For signal c, *request_update(c)* stores the request by inserting the signal into the *update_array*[5].

---

[5] The name is owed to the V2.0.1 reference implementation. For our semantics we have defined it as a set to reduce complexity in the ASM rules.

$$request\_update(c) \equiv update\_array := update\_array \cup \{c\}$$

Based on the update request within the *request update()*, the simulation scheduler performs an *update* which is outlined in the next section. Reading a signal by c.read() is expressed by a function call just returning the current value of the signal.

4.5.2 Notify

Events are implicitly generated when updating a primitive channel such as a signal. Alternatively, an event e can be explicitly sent by e.notify(). The notify operation without any parameter generates an immediate notification in the current delta cycle. All others notify operations generate events for future cycles by either delta delay or timed notification. For that, each event holds the set of all processes, which are sensitive to that event (*processes(e)*), i.e., which are resumed when the scheduler elaborates the events. Thus for immediate execution all these processes are immediately set to the status *ready*.

> **P2** :
> **if** *Self executes* $\langle$e.notify()$\rangle$
> **then** $\forall p \in processes(e) : status(p) := ready,$
> $programCounter(Self) := nextStmt(Self)$
> **endif**

Owing to the value of its parameter, the notify operation has effect on future delta cycles or time cycles. We organize the event management through the global set *pendingEvents*, which holds the set of current and future events. Therefore it is important to keep the notification time for each event as (*time(e)*). Owing to the rules of the current language reference manual, the notify operation only has effect under specific conditions. When notified, the event e is inserted in *pendingEvents* only when it is not already element of that set. Then its *time* is set to an absolute time by setting it to the specified time t plus the current time T_c. In the case that event e has already received a notification, its time is overwritten if the new time is smaller or equal than the already assigned time. That means that previous notifications on that event are cancelled when the already scheduled notification is later than the current notification request. In other words, when new notification requests are later than the already scheduled ones they are simply ignored. Finally, after executing the operation the process proceeds to the next operation by advancing the program counter.

P3 :

if *Self executes* $\langle$`e.notify(`t`)`$\rangle$

then if $e \notin pendingEvents$ **then**

    extend $EVENT$ **by** e

      $pendingEvents := pendingEvents \cup \{e\}$,

      $time(e) := T_c + t$

    endextend

    elseif $time(e) > T_c + t$ **then**

      $time(e) := T_c + t$

    endif

    $programCounter(Self) := nextStmt(Self)$

endif

SystemC V2.0 also introduces an operation to cancel events. On the execution of that operation, the event is simply removed from the existing *pendingEvents*.

P4 :

if *Self executes* $\langle$`e.cancel()`$\rangle$**then**

  $pendingEvents := pendingEvents - \{e\}$,

  $programCounter(Self) := nextStmt(Self)$

endif

4.5.3 Notify_Delayed

In addition to `notify()`, SystemC introduces a `notify_delayed()` operation. As a result of the first one the latter is available without parameters and with time specification yielding to a similar behavior. In contrast to `notify`, `notify_delayed` cannot set an immediate event on the current delta cycle, i.e., `notify_delayed` without parameters set a notification for the next delta cycle. In addition `notify_delayed` generates a run time error and simulation stops when an event e is already inserted into *pendingEvents*[6]. Except those two differences, the semantics is the same as given in rule P3. Therefore, it should be sufficient when we only give the semantics of a `notify_delayed` with time specification t here.

---

[6] Note that this is implementation specific behavior of the SystemC V2.0.1 simulator and may be subject for a modification

P5 :
if *Self executes* $\langle$`e.notify_delayed`$(t)\rangle$
then if $e \notin pendingEvents$ **then**
    extend $EVENT$ **by** e
      $pendingEvents := pendingEvents \cup \{e\},$
      $time(e) := T_c + t$
    endextend
    else $STOP$
    endif
    $programCounter(Self) := nextStmt(Self)$
endif

4.5.4   Wait

The previously outlined operations schedule events. Processes can defined to be sensitive to those events through the different variations of the `wait` and the `next_trigger` operations. Wait operations are for the synchronization of threads, `next_trigger` for methods.

For both operations we have to define the semantics of several different cases where we first distinguish static and dynamic sensitivity. It is possible to define a sensitivity list when declaring a process in the constructor of a module. That sensitivity list is denoted as the *static sensitivity* list of a process, i.e., on the notification of at least one of the events in that list, the process resumes from *suspended*. As an example, consider the following definition which declares process p in the constructor of module m (`SC_CTOR(m)`) and a sensitivity list with events $e1$ and $e2$ in the sensitivity list of p which defines that p is sensitive to $e1$ and $e2$.

```
SC_MODULE(m) {
    sc_event e1, e2;

    SC_CTOR(m) {
        SC_THREAD(p);
        sensitive << e1 << e2;
    };
};
```

The static sensitivity list is assigned for the whole lifetime of the process in the elaboration phase. Static sensitivity lists are referred to in the body of a thread by a wait operation without parameters. When executing such a wait operation, a process *Self* changes its status to *suspended* and sets its *trigger* to *staticSensitivity* which is required to identify that the process resumes based on the evaluation of the static sensitivity list. Recall that each event e holds a list

110

processes(e) in order to identify the processes which are *suspended* on those events. Therefore *Self* has to be included in *processes(e)* of all events *e* in the sensitivity list.

> **P6** :
> **if** *Self executes* ⟨wait()⟩ **then**
>   **if** *Self* ∉ *THREAD* **then** *STOP* **endif** ,
>   ∀*e* ∈ *sensitivityList(Self)* :   *processes(e)* := *processes(e)* ∪ *Self*,
>   *status(Self)* := *suspended*,
>   *trigger(Self)* := *staticSensitivity*,
>   *programCounter(Self)* := *nextStmt(Self)*
> **endif**

In the case of a wait operation with parameters the objects given in the parameters temporarily replace the ones in the static sensitivity list of the process. Because the temporary list is active only for the next invocation of the process, it is denoted as a *dynamic sensitivity list*. Dynamic sensitivity lists can either range over timeouts or events or both.

When being sensitive on a timeout *t*, the process first creates a new event and schedules this event at time $t + T_c$ by setting *time(e)*. *Self* is inserted into the list of processes of event *e* in order to be identified for later invocation when the event becomes active, i.e., when the timer expired. The event *e* is inserted into the set of *pendingEvents*. The event expression of the process is kept in *eventExpr(Self)* for later evaluation which will be outlined later. Thereafter, the process sets its *trigger* to *dynamicSensitivity* and changes *status* to *suspended*.

> **P7** :
> **if** *Self executes* ⟨wait(*t*)⟩ **then**
>   **if** *Self* ∉ *THREAD* **then** *STOP* **endif** ,
>   **extend** *EVENT* **with** *e*
>   *time(e)* := $T_c + t$,
>   *processes(e)* := {*Self*},
>   *pendingEvents* := *pendingEvents* ∪ {*e*},
>   *eventExpr(Self)* := *e*,
>   *status(Self)* := *suspended*,
>   *trigger(Self)* := *dynamicSensitivity*,
>   *programCounter(Self)* := *nextStmt(Self)*
>   **endextend**
> **endif**

We distinguish three variations of a SystemC wait operation: single event, 'and list', and 'or list'. Examples for SC_EVENTs *e*1, *e*2, *e*3 are

```
wait(e1)           // single event
wait(e1 & e2 & e3) // and list
wait(e1 | e2 | e3) // or list
```

For 'or lists' the process resumes when at least one of the given events is notified. When events are combined by &, the process resumes after all events have been notified. The 'and notification' may range over different delta cycles so that the notification history has to be managed.

We generalize all three cases and define their semantics by one ASM rule over an event expression *EventExpr*, which can be of one of the three types[7]. First, the rule sets its *trigger* to *dynamicSensitivity*. Thereafter, the process *Self* is inserted into the list of processes which have to be resumed. This is done for each event in the event expression of the wait operation. The function *eventCollection(EventExpr)* returns the set of all events used in *EventExpr*. Then the used *EventExpr* is stored since the scheduler has to access these events, e.g., for removing *Self* from all events of an or list, if one of these events has been triggered. Finally, the rule sets the process *status* to *suspended* and advances the program counter.

P8 :
if $Self$ executes $\langle \underline{\text{wait}}(EventExpr)\rangle$ **then**
  if $Self \notin THREAD$ **then** $STOP$ **endif**,
  $trigger(Self) := dynamicSensitivity,$
  $\forall e \in eventCollection(EventExpr) : processes(e) := processes(e) \cup Self,$
  $eventExpr(Self) := EventExpr,$
  $status(Self) := suspended,$
  $programCounter(Self) := nextStmt(Self)$
endif

There is the possibility to combine a timeout with an event expression. For example, we can write for time t and events $e1, e2, e3$

```
wait(t, e1 & e2 & e3)
wait(t, e1 | e2 | e3)
```

The meaning is that the process either waits for the truth of the event expression or for the timeout. We do not give an extra ASM rule here since the semantics is a generalization of event expressions towards timeouts so that this would simply need a combination the ASM rule P7 and P8.

Considering the `wait_until(expr)` operation, we do not define an extra ASM rule since it can be translated to:

```
do wait(); while (!expr);
```

---

[7] Our definition abstracts from these details and can therefore also be applied to future versions allowing the mixing of & and | operators

4.5.5   Next_Trigger

Whereas the wait operation provides a synchronization for threads, methods have the next_trigger operation for their invocation based on static and dynamic sensitivity lists. Corresponding to the different wait operations outlined in the previous section, we can distinguish exactly the same variations for the next_trigger operation:

```
next_trigger()                    // static sensitivity
next_trigger(t)                   // timeout
next_trigger(e1)                  // single event
next_trigger(e1 & e2 & e3)        // and list
next_trigger(e1 | e2 | e3)        // or list
next_trigger(t, e1 & e2 & e3)     // timeout with and list
next_trigger(t, e1 | e2 | e3)     // timeout with or list
```

The semantics of those operations is not to stop the execution of the method after executing the next_trigger rather than to continue until the last statement of the method. In the case of executing more than one next_trigger operations, only the last one determines the next trigger.

METHODs are not allowed to have a wait method suspending their execution and managing their sensitivity. Therefore we first set METHODs (see next section, Rule S2) to be static sensitive by default. If a call to next_trigger arises we remove the process from the list of suspended processes, which are kept for each event. If this call is not the first call then we have to remove this process from *processes(e)* of all dynamic events $\mathscr{E}$. Afterwards the new sensitivity is managed and the program counter is incremented. Since the definition of the next_trigger method is similar to the definition of the wait statements, we only give a very generic rule for the handling of event expressions.

P9 :
if *Self executes* ⟨**next_trigger**(*EventExpr*)⟩ **then**
  if *Self* ∉ *METHOD* **then** *STOP* **endif** ,
  if *trigger*(*Self*) = *staticSensitivity* **then**
    ∀e ∈ *sensitivityList*(*Self*) : *processes*(e) := *processes*(e) − {e}
  else
    ∀e ∈ *eventCollection*(*eventExpr*(*Self*)) : *processes*(e) := *processes*(e) − {e}
  endif
  trigger(*Self*) := *dynamicSensitivity*,
  ∀e ∈ *eventCollection*(*EventExpr*) : *processes*(e) := *processes*(e) ∪ {*Self*},
  eventExpr(*Self*) := *EventExpr*,
  programCounter(*Self*) := *nextStmt*(*Self*)
endif

---

[8] obtained by *eventCollection(eventExpr(Self))*

4.6   SystemC Scheduler

The SystemC scheduler is a separate process which we define by a separate ASM agent. It is executed as soon as all user defined processes are not executing (cf. Figure 4.1). We abbreviate this by:

$$AllProcessesNotExecuting \equiv$$
$$\forall p \in METHOD \cup THREAD \; : \; status(p) \neq executing$$

When all processes are not executing, the scheduler goes through different steps (see Fig. 4.5) by setting the ASM function *step*. Thereafter *suspended METHODs* and *THREADs* are individually resumed until they go to *suspended* again. *THREADs* change to *suspended* reaching a wait operation, *METHODs* after their last statement.

The 8 steps of the scheduler are defined in the language reference manual as follows.

Step 1: initialize
Step 2: resume a ready process
Step 3: if there are ready processes goto 2, otherwise goto 4
Step 4: update all channels with update requests
Step 5: if there are events at current time goto 2, otherwise goto 6
Step 6: if there are no more pendingEvents for future time,
        STOP simulation, otherwise goto 7
Step 7: advance the current simulation time T_c; if max simulation time
        exceeded then STOP, otherwise goto 7
Step 8: set processes ready to run for the new T_c

Note here that the language reference manual and [Grötker et al., 2002] denotes Step 1 as the initialization phase, Steps 2–3 as the evaluate phase, and Steps 4–8 as the update phase (see also Fig. 4.4 and 4.5). In our specification we take the fine grain partition into steps since we need a sufficiently detailed model of the scheduler.

The initialization of most of the ASM functions should be intuitively clear. Therefore we only sketch the initializations informally. We suppose *step* := *checkProcesses* and current time T_c to be set to 0. For a processes p with dont_initialize(), we execute an implicit wait() at the beginning, i.e., the processes become sensitive to their static sensitivity list and is *suspended*. All others processes are considered to be in status *executing* and immediately start their execution.

For each clock, we introduce one extra SystemC module with one thread as it is shown in Figure 4.6. Since these modules also inherit from the interface class 'sc_signal_in_if<bool>' (multiple C++-inheritance), they behave like a combination of input signals and modules, i.e., they can be used wherever input signals are used.

```
clock() {
  bool value = !posedge_first;
  write(value);
  wait ( start_time );
  while (1) {
    value = !value;
    write(value); // this generates an
                  // update_request in all cases
    if (value) wait( period * duty_cycle );
    else    wait( period * (1-duty_cycle) );
  }
}
```

Figure 4.6. SystemC clock behavior

All other ASM functions are assumed to be *undef* and sets and lists to be empty.

After initialization the *Scheduler_ASMmodule* agent executes as follows. Compared to the previously introduced steps of the scheduler, owing to technical ASM specific matters we have to combine Step 4–8 to two ASM rules, which check for events and either stop the simulation or advance the time. In order to keep the ASM specification readable, we use ASM macros as placeholders for the individual sequential steps *ResumeProcess*, *CheckReadyToRun*, *UpdateChannels*, *CheckEvents*, and *AdvanceTime*, so that we finally arrive at the following definition.

S1 :
if *AllProcessesNotExecuting* **then**
   ResumeProcess,
   CheckReadyToRun,
   UpdateChannels,
   CheckEvents,
   AdvanceTime
endif

The first step is defined by *ResumeProcess* and checks for all $p \in PROCESS$ with status *ready*. As defined by SystemC, only one process is arbitrarily selected from the set of processes fulfilling that condition. When there exists such a process p, p is resumed by setting its status to *executing*. Additionally, the *trigger* of each METHOD is reset to *staticSensitivity*. Thereafter the scheduler proceeds to the next step by setting *step* to *checkReadyToRun*. Note that the scheduler does not immediately execute the next step since the process first has to execute, i.e., the condition *AllProcessesExecuting* first has to become true again.

S2 :
$ResumeProcess \equiv$
if $step = resumeProcess$ **then**
choose p **in** $PROCESS$ **satisfying** $status(p) = ready$
$\quad status(p) := executing,$
$\quad$**if** $p \in METHOD$ **then**
$\quad\quad trigger(p) := staticSensitivity$
$\quad\quad \forall e \in staticSensitivity(p) : processes(e) := processes(e) \cup \{p\}$
$\quad$**endif**
endchoose
$step := checkReadyToRun$
endif

The next step given by *CheckReadyToRun* checks if there are more processes *ready* to run. If there is at least one the scheduler goes back to resume it. Otherwise the scheduler proceeds to update the channels.

S3 :
$CheckReadyToRun \equiv$
if $step = checkReadyToRun$ **then**
$\quad$**if** $\exists p \in PROCESS : status(p) := ready$ **then**
$\quad\quad step := resumeProcess$
$\quad$**else**
$\quad\quad step := updateChannels$
$\quad$**endif**
endif

In the third step *UpdateChannels* takes all update requests which are held in the *update_array* and perform an *update()* on each of them.

S4 :
$UpdateChannels \equiv$
if $step = updateChannels$ **then**
$\quad \forall c \in update\_array : update(c),$
$\quad update\_array := \emptyset,$
$\quad step := checkEvents$
endif

The *update* function is different for each channel type. For signals the update is defined by the following ASM macro definition. Only in the case when the current value of the signal is different from the new requested value, it is updated by its new value and a corresponding event (*default event*) is inserted into the global set of pendingEvents. For that event its time has to be set to the current

116

time T_c since the processes suspended on the signal update have to be triggered for the next delta cycle at current time. Note here that those two additional events for Boolean signals (one notifying the positive and one notifying the negative edge) shall be inserted at this point. We leave this unspecified since it would introduce unnecessary complexity.

S5 :
$update(c) \equiv$
if $new\_value(c) \neq value(c)$ then
  $value(c) := new\_value(c),$
  $time(default\_event(c)) := T_c,$
  $pendingEvents := pendingEvents \cup \{default\_event(c)\}$
endif

After the step *UpdateChannels* the scheduler checks for events by *Check-Events*. If there are no further events, i.e., *pendingEvents* is empty, the simulation stops. Otherwise if there are pending events at the current time, the corresponding processes are triggered and set to *ready* which is defined within the macro *TriggerProcesses()*. Thereafter the scheduler proceeds to resume those. In the case that there are no events at the current time, the scheduler proceeds to *advanceTime*.

S6 :
$CheckEvents \equiv$
if $step = checkEvents$
  then if $pendingEvents = \emptyset$ then $STOP$
    elseif $\exists e \in pendingEvents : time(e) = T_c$ then
      $TriggerProcesses(T_c),$
      $step := resumeProcess$
    Else $step := advanceTime$
    endif
  endif
endif

The macro *TriggerProcesses* is used for setting processes to *ready* for T_c and for the next time point T_n on the next page in Rule S10. For triggering processes at time t, we first have to identify all events e whose *time(e)* match the given time t. Additionally we identify all processes within the lists of those events. Thereafter we distinguish the processes, which suspended on the static and dynamic sensitivity lists since we have to separately define both cases, which are specified by the macros *TriggerStatic* and *TriggerDynamic*. In both cases the event has to be removed from *pendingEvents* afterwards.

S7 :
$TriggerProcesses(t) \equiv$
var p **ranges over** $PROCESS$
var e **ranges over** $EVENT$
if $e \in pendingEvents \wedge time(e) = t$ **then**
  if $p \in processes(e)$ **then**
    if $trigger(p) = staticSensitivity$ **then**
      $TriggerStatic(p, e)$
    endif
    if $trigger(p) = dynamicSensitivity \wedge$
      $e \in eventCollection(eventExpr(p))$ **then**
      $TriggerDynamic(p, e)$
    endif
    $pendingEvents := pendingEvents - \{e\}$
  endif
endif

In the case of static sensitivity we check if the event is in the sensitivity list of the process before setting it to *ready* and deleting the process from all the process lists of all events in its sensitivity list.

S8 :
$TriggerStatic(p, e) \equiv$
if $e \in sensitivityList(p)$ **then**
  $status(p) := ready,$
  $\forall e \in sensitivityList(p) : processes(e) := processes(e) - \{p\}$
endif

In the case of a dynamic sensitivity we evaluate the event expression. When returning true the *status* of the corresponding process can be set to *ready*. To keep it short we only give an informal definition of the function *eval* here since its semantics should be intuitively clear and we want to keep it flexible enough for future SystemC extensions. The function *eval()* takes the event expression over the dynamic sensitivity as a parameter. With an event 'or list' expression and combinations with timeouts, *eval* becomes true when one of the events or the timeout is true in the current cycle. When ranging over an 'and list' event expression, the history of the events has to be kept over past simulation cycles and the events have to be marked as true in the event list after they become true. As soon as all events are marked true, *eval* becomes true. In those cases the corresponding process is set to *ready*. Finally, the process has to be removed from all the process lists of all events in the *eventCollection* of the given event expression and the *eventExpr* of the process is set to *undef*.

$$S9 :$$
$$TriggerDynamic(p,e) \equiv$$
if $eval(eventExpr(p))$ **then**
  $status(p) := ready,$
  $\forall e \in eventCollection(eventExpr(p)) :$
    $processes(e) := processes(e) - \{p\},$
  $eventExpr(p) := undef$
endif

In *AdvanceTime* the current simulation time T_c is updated by the next time T_n and processes are triggered. For triggering the processes we take the same macro as defined before with T_n as a parameter.

$$S10 :$$
$$AdvanceTime \equiv$$
if $step = advanceTime$ **then**
  $TriggerProcesses(T_n),$
  $T_c := T_n,$
  $step := resumeProcess$
endif

T_n is computed by considering the minimum of all times of events in the set of *pendingEvents* that are greater than T_c.

$$T_n = round\_off(min\{t \mid t = time(e)\}, \texttt{sc\_get\_time\_resolution()}),$$
$$where\ e \in pendingEvents\ and\ time(e) > T_c$$

The minumum is rounded off with respect to time resolution set by the user which is given by SystemC function sc_get_time_resolution(). When not explicitly modified, the default time resolution is nanoseconds.

4.7 Example

This section gives a detailed insight into the execution of our ASM model of the SystemC V2.0 simulator on the ASM virtual machine. We execute a simple SystemC bit counter on the defined ASMs agents. The counter was originally introduced in [Delgado Kloos and Breuer, 1995] where it served as a running VHDL example for all articles in the book. There it was also used to outline the ASM based semantics of the VHDL'93 simulator. We give the corresponding SystemC code in Fig. 4.7. Fig. 4.8 additionally sketches the structure of the different modules, signals, ports, and outlines their connections. We presume that the reader has basic knowledge of SystemC V2.0. Otherwise we refer to the SystemC introduction given in [Grötker et al., 2002].

In the example sc_main defines processes p3.assign as an instance of SystemC module m2 and clk as an instance of clock (Fig. 4.8). clk is defined

```
SC_MODULE(m1) {                       SC_MODULE(m2) {
  sc_in<bool> X;                        sc_in<bool>        X;
  sc_out<bool> Y;                       sc_out<sc_int<2>> Y;
                                        sc_signal< bool > s0;
  void behaviour() {                    sc_signal< bool > s1;
    while (1) {                         m1 *p1, *p2;
      do {
        wait();                         void assign() {
      } while(X.read()==true);            Y.write(((sc_int<0>)s0,
      wait(1, SC_NS);                               (sc_int<1>)s1));
      Y.write(true);                    };

      do {                              SC_CTOR(m2) {
        wait();                           SC_METHOD(assign);
      } while(X.read()==true);            sensitive << s0 << s1;
      wait(1, SC_NS);                     p1 = new m1("p1");
      Y.write(false);                     p2 = new m1("p2");
    }                                     p1->X(X);
  };                                      p1->Y(s0);
                                          p2->X(s0);
  SC_CTOR(m1) {                           p2->Y(s1);
    SC_THREAD(behaviour);               };
    sensitive << X;                   };
    dont_initialize();
  };
};


          int sc_main(int argc, char** argv) {
            sc_signal<sc_int<2> > y;
            m2 *p3 = new m2("p3");

            sc_clock clk("clk",10,SC_NS,0.5,5,SC_NS,true);

            p3->X(clk.signal());
            p3->Y(y);
            sc_start();
          }
```

Figure 4.7. SystemC example

with a period of 10 *ns*, a duty cycle of 50 percent (the percentage where the clk is true), and a start time of 0 *ns*. The last clk parameter defines that the first edge at the start time is a negative one. Recall that in SystemC a clock is a subclass of a signal and a module, so that a clock can be considered as a thread, which determines its own value. In the following execution we refer to the code given in Figure 4.6 on Page 114 which defines the clock's behavior by the means of a SC_THREAD. Note also that m2 and m1 both have a static sensitivity list, just after the declaration of the corresponding method and thread, respectively. The clock clk triggers the p3 module via the X port which directly connects to the port X of p1. The outputs of p1 and p2 are connected to signals s0 and s1 which both combine to the value of output port Y which directly connects to signal y in sc_main.

When compiling and executing the given SystemC program, SystemC first elaborates the module structure and instantiates and connects p1, p2, p3, and clk as described before. For the initialization we have to instantiate one ASM agent as an instance of the ASM *SCHEDULER_ASMmodule* (cf. Section 4.6) and four agents of the ASM *PROCESS_ASMmodule* (cf. Section 4.5), one for each user defined process (p3.assign, p1.behaviour, p2.behaviour) and an extra one for the clock clk.

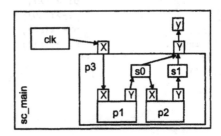

Figure 4.8.   SystemC example — structural view

Table 4.1 gives an overview of the waveform of all signals over the considered simulation time, from 0 *ns* to 22 *ns*. It shows that clk starts with *false* and toggles every 5 *ns* between t (true) and f (false). The last row gives the integer value of the primary counter output where s0 and s1 are bit components represented as *bool*.

4.7.1   Initialization

After elaboration of the program, i.e., executing it from the from the beginning of sc_main to the call of sc_start(), the simulation scheduler is invoked starting with the initialization phase. Owing to the rules of SystemC initialization, we also initialize our ASM functions. If not stated otherwise, we set

Signal/Time	0–4 ns	5–9 ns	10–14 ns	15–19 ns	20–22 ns
clk	f,f,f,f,f	t,t,t,t,t	f,f,f,f,f	t,t,t,t,t	f,f,f
s1	f,f,f,f,f	f,f,f,f,f	f,f,f,f,f	f,f,f,f,f	f,f,t
s0	f,f,f,f,f	f,f,f,f,f	f,t,t,t,t	t,t,t,t,t	t,f,f
y	0,0,0,0,0	0,0,0,0,0	0,1,1,1,1	1,1,1,1,1	1,0,2

Table 4.1.

functions to *undef* and assume lists and sets to be $\emptyset$. The simulation time resolution and the default simulation time unit are set to their default values, i.e., SC_PS and SC_NS, respectively. The current simulation time T_c is initialized by the value 0.

The universe *PROCESS* has three distinguished elements: *p1.behaviour*, *p2.behaviour*, *p3.assign*, and *clk.clock*. Since all module instances have only one process, we abbreviate the processes by *p1, p2, p3*, and *clk* in the remainder of this section. The domain *EVENT* is initialized with four default events, one for each signal including the one for the clock:

EVENT	signal	time
e_1	*clk*	*undef*
e_2	*s0*	*undef*
e_3	*s1*	*undef*
e_4	*y*	*undef*

For each signal we initialize the ASM function *event* with the corresponding default event in order to retrieve events for signals, e.g., e_1 is the default event of signal *clk*. Additionally we set *sensitivityList* for all processes as given in Table 4.2 and assume all *triggers* to be of *staticSensitivity*.

Since dont_initialize() is set, the *status* of *p1* and *p2* both become *suspended* and both are inserted into the lists *processes*(e_1) and *processes*(e_2). Additionally their *programm_counters* are assigned to their first statement and set to the beginning of the while loop. The *status* of the other processes, *clk* and *p3*, are both set to *executing*. Thereafter *clk* starts executing an instance of the code given in Fig. 4.6, i.e., it initializes *val* by *false*, executes write(val), and suspends on a wait(start_time) with *dynamicSensitity*. The write statement does not schedule an update request since the new value equals the initial value of the clock. The wait statement schedules a timeout by inserting e_5 with *time* 5 and *processes*(e_5) = {*clk*} into *pendingEvent*. Furthermore, *eventExpr(clk)* is set to *e* by the wait statement. *p3* executes Y.write() upon which no update request is set since the current value of *Y* equals its new value. *p3* suspends after the last statement with *trigger=staticSensitivity* so that at the end of the initialization we arrive at a list with one pending event.

pendingEvents	processes	time
e_5	clk	5 ns

The list of four processes and their values after the initialization is summarized in Table 4.2, where the sensitivity list of each process is determined by the (default) events associated to the signals of the processes' sensitivity, or signals reachable via the ports in the processes' sensitivity with respect to the previously introduced table of their default events e_1-e_4.

PROCESS	status	trigger	sensitivityList	programCounter
clk	suspended	*dynamicSensitivity*		`wait(start_time);`
p3	suspended	*staticSensitivity*	e_2, e_3	`Y.write()`
p1	suspended	*staticSensitivity*	e_1	`while(1){`
p2	suspended	*staticSensitivity*	e_2	`while(1){`

Table 4.2. Status of processes after initialization

After initialization the scheduler runs iteratively through *evaluate* and *update* phases and advances delta and time cycles as given in Fig. 4.4 and Fig. 4.5. The remainder of this section sketches the individual steps of the simulation scheduler with respect to the simulation time.

4.7.2 Time 0 *ns*

Evaluate. In the first simulation cycle the scheduler starts with the evaluate phase by executing *ResumeProcess* (Rule S2) and *CheckReadyToRun* (Rule S3). Since there are no processes with *status ready*, it immediately proceeds to the *update* phase and to *UpdateChannels* (Rule S4).

Update. Since there are no update requests and no events at the current simulation time, the scheduler advances through the steps *CheckReadyToRun*, *UpdateChannels*, *CheckEvents* and *AdvanceTime* (Rules S2–S10). In *AdvanceTime* (Rule S10)it sets the current time T_c to the next greater time in the list of pending events, i.e., $time(e_5) = 5$ *ns*. The final *TriggerProcesses* in Rule S10 considers process *clk* to be *ready* since $time(e_5) = 5$ *ns* $= T_n$ and $processes(e_5) = \{clk\}$. Owing to the dynamic sensitivity of *clk* and since $e_5 \in eventCollection$ of *eventExpr(clk)* it performs *TriggerDynamic(clk,e_5)* upon which the *status* of *clk* is set to *ready*.

4.7.3 Time 5 *ns*

Evaluate. After advancing T_c to 5 *ns*, the scheduler returns to *ResumeProcess* where *clk* is invoked as a *ready* process. *clk* first enters the while loop and inverts its value (to *true*) and then assigns it by executing the write statement. The latter schedules an update request, i.e., *clk* is inserted into the global *update_array* since the new value is different from the current one. Thereafter *clk* suspends on `wait(10 * 0.5)` which schedules e_6 at $T_c + 5$ *ns* $= 10$ *ns*.

pendingEvents	processes	time
e_6	*clk*	10 *ns*

Update. Thereafter the scheduler goes to *UpdateChannels* and detects the update request on *clk* so that it inserts its (default) event e_1 with time T_c because *clk* connects to *p1* via its input ports (see Fig. 4.8). The list of pending events finally shows as follows.

pendingEvents	processes	time
e_6	*clk*	10 *ns*
e_1	*p1*	5 *ns*

In the next step *CheckEvent* checks for events at the current time and thus matches e_1. As a result the *status* of $p1 \in processes(e_1)$ is set to *ready* by *TriggerStatic*. This macro also removes *p1* from *processes(e_1)*.

Evaluate. After returning to *ResumeProcess* we identify one *ready* process. *p1* is set to *executing* and finally suspends on its first wait statement with *trigger = staticSensitivity*. Now *p1* is inserted into *processes(e_1)*.

Update. Since no more processes are *ready* no update requests are available, and no events are pending for T_c, the scheduler goes through several steps and finally advances the simulation time to 10 *ns* which is given by e_6.

4.7.4 Time 10 *ns*

Evaluate. After advancing T_c to 10 *ns*, the scheduler sets *clk* to *executing* in *ResumeProcess*, which inverts its value to *false* and performs a write operation which executes an update request on *clk* before it suspends with a *5 ns* timeout when executing the wait operation. The first one schedules event e_1 at 10 *ns*, the latter schedules e_7 at 15 *ns*.

Update. Thereafter *UpdateChannels* evaluates the update requests on *clk*, which results in scheduling e_1 for 10 *ns* for triggering the process *p1*.

pendingEvents	processes	time
e_7	*clk*	15 *ns*
e_1	*p1*	10 *ns*

CheckEvents checks for events at the current simulation time. In *Trigger-Static* of *TriggerProcesses*, *p1* is set to *ready* since e_1 is in its sensitivity list, i.e., *processes(e_1)* = {*p1*}.

Evaluate. *p1* starts executing in step *ResumeProcess* and exits the first loop since X.read() becomes *false* and suspends on wait(1, SC_NS) scheduling event e_8 for 11 *ns*. At this time *processes(e_1)* is empty and $p1 \in processes(e_8)$.

pendingEvents	processes	time
e_7	*clk*	15 *ns*
e_8	*p1*	11 *ns*

Update. No more update requests and events are available at the current time so that T_c is advanced to 11 *ns* given by *time(e₈)* and *p1* becomes *ready*.

4.7.5 Time 11 *ns*

Evaluate. At time 11 *ns*, *p1* becomes *executing* in *ResumeProcess*. It writes a *true* to *s0* via its output port. After inserting *s0* into the *update array* it suspends with static sensitivity on the wait statement in the second while loop. This includes *p1* again into *processes(e₁)*.

Update. In *UpdateChannels* the update request schedules the (default) event of *s0* with *processes p2* and *p3* since $e_2 = default\_event(s0)$ is in the sensitivity lists of both.

pendingEvents	processes	time
e_8	clk	15 ns
e_2	p2, p3	11 ns

Evaluate. When returning to *ResumeProcess*, *p2* executes and advances to the first wait statement which suspends *p2* based on static sensitivity. *p3* executes and schedules an update request for *y* since its current value 0 does not equal the new value 1.

Update. In *UpdateChannels* the value of *y* is set to 1. This would also schedule an event for T_c. Since *y* is not in the sensitivity list of any process it can be ignored. No other events are available at the current time, so that T_c is advanced to 15 *ns* and the status of *clk* becomes *ready*.

4.7.6 Time 15 *ns*

Evaluate. The next cycle starts with *ResumeProcess*, which executes *clk*. *clk* sets is value to *true* and changes the value by the write operation yielding an update request. Thereafter it suspends with a timeout generating event e_9 for 20 *ns*.

Update. In *UpdateChannels* the update request executes an update on *clk* and schedules the event e_1 for process *p1* at T_c. *CheckEvents* finally sets *p1* to *ready*.

pendingEvents	processes	time
e_9	clk	20 ns
e_1	p1	15 ns

Evaluate. Returning to *ResumeProcess*, the *ready* process is executed. When executing *p1* it suspends on the next wait in the second while loop with static sensitivity since the value of *clk*, which is connected to input port *X* of *p1* is *true*.

Update. After proceeding to *AdvanceTime*, the scheduler advances T_c to 20 *ns* and *status(clk)* becomes *ready*.

4.7.7   Time 20 *ns*

Evaluate. *ResumeProcess* sets *clk* to *executing* which changes its value to *false* with an update request on *clk* before it suspends on a wait and schedules event e_{10} for 25 *ns*.

Update. In *UpdateChannels* the update request on *clk* sets the process *p1* to be *ready* since it is $\in processes(e_1)$.

Evaluate. When returning to the step *ResumeProcess* the *ready* process resumes. *p1* exits the loop since X.read() returns *false*. Next it suspends at wait(1,SC_NS) and generates event e_{11} for time 21 *ns*. The set of pending events now has two entries.

pendingEvents	processes	time
e_{10}	*clk*	25 *ns*
e_{11}	*p1*	21 *ns*

Update. Since there are neither current update requests nor pending events at T_c, the scheduler advances the time to 21 *ns* so that *p1* is set to *ready* since it is associated with event e_{11}.

4.7.8   Time 21 *ns*

Evaluate. *p1* is set to *executing* which assigns *false* to *s0* via its output port. This generates an update request since *s0* changes from *true* to *false*. Thereafter *p1* returns to the first while loop and suspends on the first wait statement with dynamic sensitivity.

Update. In *UpdateChannels* the update request causes an update of signal *s0* and schedules event e_2 for processes *p3* and *p2* at time 21 *ns*.

pendingEvents	processes	time
e_{11}	*clk*	25 *ns*
e_2	*p2,p3*	21 *ns*

Evaluate. *ResumeProcess* executes *p2* and proceeds to wait(1,SC_NS) since the input port connecting to *s0* became *false*. Suspending on the wait schedules e_{12} for 22 *ns*. *p3* just executes the write statement and generates an update request on y.

Update. In *UpdateChannels* the given update request is evaluated. Since *s0* and *s1* both are *false*, the update sets the value of y to 0. No event is scheduled since *default_event(y)* is not in the sensitivity list of any process.

pendingEvents	processes	time
e_{11}	*clk*	25 *ns*
e_{12}	*p2*	22 *ns*

Finally, the scheduler advances to *AdvanceTime* and increments the simulation time to 22 *ns*, owing e_{12}, so that *p2* becomes *ready* again.

4.7.9 Time 22 *ns*

Evaluate. At the beginning of this simulation cycle, *p2* executes and assigns *true* to $s1$ via its output port which generates an update request on $s1$ before suspending on the wait operation in the next while loop.

Update. Thereafter the update request updates $s1$ to *true* and schedules the (default) event of $s1$ (e_3) with $processes(e_3)=\{p3\}$. The next step thus sets *p3* to *ready* yielding *p3* to resume and execute.

Evaluate. $p3$ is resumed and writes to Y which generates an update request on y with new value 2. Note here that the value of y is composed of two bit components $s0=false$ and $s1=true$. The value of y is finally updated to 2 by *UpdateChannels*. We can skip the scheduling of the corresponding event since *default_event(y)* is not in the sensitivity list of any process. Thus *AdvanceTime* determines to advance T_c to 25 *ns* which completes our investigations.

4.8 Conclusions

This article introduces the formal definition of the SystemC V2.0 simulation semantics through the specification SystemC specific statements and their interaction with the simulation scheduler by means of ASM rules based on the formal ASM theory. The definition is fully compliant with the notion given in the SystemC manual and presents an abstract model of the SystemC V2.0.1 reference implementation. It is a formal but intuitive description of the SystemC V2.0, which may complement the SystemC language reference manual. Since the definition given in this chapter compares to the patterns of the ASM definition of VHDL'93 [Börger et al., 1995] and SpecC [Müller et al., 2002], our formal definition provides a suitable basis to study interoperabilities with VHDL'93 and SpecC.

Acknowledgments

The work described herein was partly funded by the German Research Council (DFG) within the priority program *Integration of Specification Techniques with Engineering Applications* under grants number KR 1869/3-2 and within the DFG Research Center 614 (*Self-Optimizing Systems of Mechanical Engineering*). We also appreciate the valuable comments of Uwe Glässer who reviewed this article.

Chapter 5

SystemC as a Complete Design and Validation Environment

Alessandro Fin, Franco Fummi, Graziano Pravadelli

Dipartimento di Informatica, Università di Verona, Italy

Abstract   Synthesis tools for SystemC descriptions are mature enough to cover the design flow from the system level to the gate level, whilst SystemC centered validation methodologies are still under development. This chapter presents a complete validation framework for SystemC designs based on a mix of functional testing and model checking. It is based on a fault model and a test generation strategy that are applicable through the whole design flow from system level to gate level. This is achieved by exploiting the SystemC 2.0 simulation efficiency and by performing an effective test patterns inheritance procedure. Fault simulation and automatic test pattern generation (ATPG) have been efficiently integrated into a unique C++ executable code linked to the SystemC model. In the case of mixed SystemC/HDL designs co-simulation is avoided by using a HDL to SystemC automatic translator, which produces a uniform executable SystemC model. Perturbed (faulty) SystemC descriptions are generated by injecting high-level faults into the code that are detected by using an ATPG. Undetected faults may be either faults hard to detect or design errors, thus they are further investigated by using model checking on the synthesized SystemC code. In this way the intrinsic characteristic of SystemC 2.0, in order to cover the whole design flow, is further extended by producing a SystemC-based validation environment that links a SystemC model with tools aiming at identifying design errors at all abstraction levels.

5.1   Introduction

The SystemC language has become a new standard in the EDA field and many designers have started to use it to model complex systems. SystemC has been mainly adopted to define abstract models of hardware/software components, since they can be easily integrated for rapid prototyping. However, it can also be used to describe modules at a higher level of detail, e.g., RT level hardware descriptions and assembly software modules. Thus it would be possible to imagine a SystemC based design flow [Fin et al., 2001a], in which the

W. Müller et al. (eds.),
SystemC: Methodologies and Applications, 127–156.

system description is translated from one abstraction level to the following one by always using SystemC representations. The adoption of a SystemC based design flow would be particularly efficient for validation purpose. In fact, it allows one to define a homogeneous functional validation procedure, applicable to all design phases, based on the same fault model and on the same test generation strategy.

Testing SystemC descriptions is still an open issue, since the language is new and researchers are looking for efficient fault models and coverage metrics, which can be applied indifferently to hardware and software modules. Previous automated approaches to functional test generation have targeted statement coverage [Malaiya et al., 1995, Corno et al., 1997], i.e., coverage of the statements in a VHDL or C design description, or coverage of transitions in a finite state machine model of a design [Cheng and Krishnakumar, 1996]. An automated approach was used to obtain functional tests that targeted both statement coverage and path coverage [Yu et al., 2002]. However, very few approaches have been presented in the literature directly targeting SystemC code [Ferrandi et al., 2002b, Harris and Zhang, 2001] and they are all focused on RTL characteristics of the language.

The proposed validation methodology tries to cover the whole design flow of a digital system from the system level of description to its structural representation by using the same fault model and test generation technique. The accurate description level and simulation performance reached by a SystemC 2.0 description allow the definition of an efficient validation procedure. In fact, exploiting SystemC simulation performance, fault simulation, and automatic test pattern generation can be directly integrated into a unique C++ executable code. On the contrary, they are very time consuming activities by considering traditional hardware description languages.

The chapter is organized as follows. Section 5.2 explains the overall methodology. Section 5.3 describes how design errors can be modeled by using a high-level fault model. Section 5.4 describes the proposed validation environment based on the combination of three different well known validation techniques (fault simulation, ATPG and model checking) implemented in SystemC. Section 5.5 presents some approaches to perform efficient fault simulation. Experimental results are showed in Section 5.6, whilst Section 5.7 is devoted to concluding remarks.

5.2 Methodology Overview

The proposed methodology shows how SystemC 2.0 features have been used to create an efficient framework for the validation of digital systems. The steps of the methodology can be summarized as follows:

- **Design Error Modeling.** The aim of the methodology is the identification of design errors (over-specified functionality, wrong modules interconnections, etc.). They are modeled by using the *bit coverage* fault model, which has been proved in [Ferrandi et al., 2002a] being related to this kind of errors. In this chapter bit coverage is extended to deal with the new features of SystemC 2.0. The automatic tool [Fin et al., 2001b] is used to generate the fault list and the perturbed descriptions of the design under validation with respect to the bit coverage criterion. The faulty and fault-free descriptions are then used during the high-level validation process;

- **High Level Validation.** A functional fault simulator (*FS*) and automatic test pattern generator (*ATPG*) have been developed adopting the bit coverage fault model. The faulty and fault-free descriptions are compared by the *FS* and *ATPG* in order to detect as many faults as possible. Moreover, the fault simulator can be used to re-simulate the same test patterns at different levels of abstraction. Undetected faults may be simply hard to detect or redundant faults representing design errors, thus they are further investigated by using model checking as partially proposed in [Fin et al., 2002];

- **Efficient High Level Fault Simulation.** More and more SystemC code is used to model and simulate digital designs, however, a large number of designs are still based on more traditional hardware description languages, such as VHDL or Verilog. Thus techniques for efficiently integrating these languages with SystemC are a hot topic spreading from *modules* to *EDA tools porting*. This chapter explains different alternatives to address the SystemC vs. VHDL integration problem with respect to *FS* and *ATPG*.

5.3   Design Error Modeling

The aim of the proposed validation strategy is to detect design errors. However, the adopted fault model allows to unify in the same approach testability estimation and physical faults detection too. A fault model is necessary to simulate the effects of errors on the *design under verification* (DUV) and to estimate its testability. In this section general aspects concerning fault models are presented and the application of bit coverage to SystemC 2.0 is deeply explained.

5.3.1   Design Errors Characterization

Many kinds of design errors can be included into the high level description of a digital system. Functionality may be over-specified and modules can be inadequately interconnected. Moreover, some parts of the code can be unreachable because of incorrect formulation of conditional statements. The aim of high level fault models is to simulate the behavior of the DUV in presence of such errors. In the literature many studies are related to the identification of a widely accepted high level fault model. Functional testing considers mainly two classes of fault models: functional fault models [Abraham and Fuchs, 1986, Abraham and Agarwal, 1985], which are based on the input/output relationship of higher level primitives, and behavioral fault models [Gosh, 1988, Gosh and Chakraborty, 1991, Chakraborty and Gosh, 1988], which are described on the top of the procedural description of the DUV.

The adoption of a fault model is closely related to the technique used for injecting faults into the DUV. The logic values of the simulated design must be modified by considering the alterations imposed by the faults. In this way ATPG can be applied to detect as many injected faults as possible. The achieved fault coverage highlights both the presence of design errors and testability problems.

Figure 5.1 shows the accurate taxonomy for simulated fault injection techniques discussed in [Gracia et al., 2001]. Fault injection techniques based on simulator commands have been presented in [Jenn et al., 1994]. Simulator commands are exploited to modify the value of the model signals and variables without altering code description. Only simple fault models can be implemented by using simulator commands, as their applicability depends on the set of commands supplied by the simulator. On the other hand, code modification techniques are based on saboteurs [Jenn et al., 1994] or on mutants of the model components [Armstrong et al., 1992]. A saboteur is an artificial HDL component added to the original design, which does not affect the DUV behavior during the normal operation. The goal of a saboteur is to alter the properties of the target signal (e.g., value, timing response), when the corresponding fault is injected. Instead a mutant is a component functionally equivalent to the replaced one, but it has an additional functionality which allows one to simulate the faulty behavior owed to the injected fault. Mutants can affect the values of the internal variables and signals or they can compromise the model timing. Both saboteurs and mutants require code modifications. In some cases a new description of the DUV must be generated for each modeled fault. Other techniques are based on extensions of the HDL with specific data types and signals [Sieh et al., 1997]. The main disadvantage of these approaches concerns the need of *ad hoc* HDL compilers and control algorithms to manage the language extensions.

5.3.2   Fault Model

The fault model adopted in the proposed methodology is the *bit coverage*, which simulates under the single fault assumption:

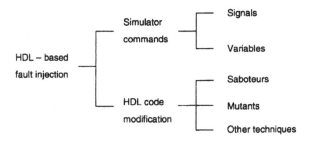

Figure 5.1. A simulated fault injection taxonomy

- **Bit failures**. Each variable, constant, signal or port is considered as a vector of bits. Each bit can be stuck at zero or one;

- **Condition failures**. Each condition can be stuck at true or stuck at false, thus removing some execution paths in the faulty SystemC representation.

The fault model excludes explicitly the incorrect behavior of the elementary operators (e.g., $+$, $-$, $*$, etc.). Therefore only single bit input or output faults are modeled, including all operator's equivalent faults.

```
//fault free code          //faulty code
data.write(req_data);       req_data[0] = '1'; //forced stuck at 1 bit
                            data.write(req_data);

                            req_data[1] = '1';
                            data.write(req_data);

                            ... ...

                            req_data[7] = '1';
                            data.write(req_data);
```

Figure 5.2. Effect of stuck at 1 bit failure application on variable req_data

Figure 5.2 and Figure 5.3 show respectively how a bit fault and a condition fault modify the description of a SystemC DUV. The bit coverage fault model directly covers all language characteristics of SystemC 1.2, which models RTL descriptions. On the contrary, it must be extended to cover the new SystemC 2.0 language features concerning events and channels.

- **Events** are faulted in two different ways: by changing the eventual parameter of the notify method and by avoiding the event notification. Pa-

132

```
//fault free code                              //faulty code
if(opcode.read()=="00110101"){                 pc = pc + 1;
  pc = pc + 1;                                  out_old = out.read();
  out_old = out.read();                         out.write(pc);
  out.write(pc);
} else {
  out.write(out_old);
}
```

Figure 5.3. Effect of stuck at true condition failure application

rameter modification is already modeled by the bit coverage fault model, since the parameter is a sc_time variable or constant. On the other hand, notification avoidance is obtained inserting an extra conditional statement. Figure 5.4 shows how an event is faulted. Parameter modification is performed by mean of a faulty value for t. Next Section shows the technique for injecting faulty elements and it deeply describes this point.

- **Channels** are similar to port and signals from the fault model point of view. They can be faulted by modifying data managed by channels methods. Such data are faulted by following the bit failure strategy of the bit coverage fault model.

```
//fault-free code                          //faulty code
void write(char c, sc_time t) {            void write(char c, sc_time t) {

  wait(read_event);                          wait(read_event);

  data = c;                                  data = c;
  ++ num_elements;                           ++ num_elements;
                                             if !(avoidance_fault)
  write_event.notify(t);                     write_event.notify(faulty_t));
}                                          }
```

Figure 5.4. Faulty event notification

Bit coverage can be easily related to the other metrics, developed in the software engineering field [Myers, 1999] and commonly used in functional testing.

- **Statement coverage**. Any statement manipulates at least one variable or signal. The bit failures are injected into all variables and signals on the left hand and right hand side of each assignment. Thus at least one test vector is generated for all statements. To reduce the proposed fault model to statement coverage it is thus sufficient to inject only one bit failure into one of the variables (signals) composing a statement. In conclusion, the bit coverage metric induces an ATPG to produce a larger number of test patterns with respect to statement coverage and it guarantees to cover all statements.

- **Branch coverage.** The branch coverage metric implies the identification of patterns which verify the execution of both the true and false (if present) paths of each branch. Modeling of our condition failures implies the identification of patterns which differentiate the true behavior of a branch from the false behavior, and vice versa. This differentiation is performed by making stuck at true (false) the branch condition and by finding patterns executing the false (true) branch, thus executing both paths. In conclusion, the proposed bit coverage metric includes the branch-coverage metric.

- **Condition coverage.** The proposed fault model includes condition failures which make stuck at true or stuck at false any condition disregarding the stuck at values of its components.

- **Path coverage.** The verification of all paths of a SystemC method can be a very complex task owing to the possible exponential grow of the number of paths. The proposed fault model selects a finite subset of all paths to be covered. The subset of covered paths is composed of all paths that are examined to activate and propagate the injected faults from the inputs to the outputs of the SystemC design module within a given time limit.

- **Block coverage.** In [Corno et al., 1997] statement coverage has been extended by partitioning the code in blocks and by activating these blocks a fixed number of times. This block coverage criterion is included in the proposed fault model in the case the number of bit faults included in a block is larger than the number of times the block is activated. In fact, a test pattern is generated for each bit fault, thus the block including the fault is activated when the fault is detected.

In conclusion, the bit coverage metric, applied to SystemC descriptions, unifies into a single metric the well known metrics concerning statements, branches and conditions coverage; an important part of all paths is also covered and all blocks of a description are activated several times. Moreover, bit coverage has been chosen since there is a high correlation between stuck at faults at different levels of abstraction [Ferrandi et al., 2002a] and this aspect can be exploited to design an incremental testing procedure as stated in [Fin et al., 2001a].

5.3.3 SystemC Faulty Description Generation

According to the bit coverage fault model SystemC descriptions are automatically modified by injecting modeled faults. The injection technique has been developed as part of the testing framework AMLETO [Fin et al., 2001b] and it is based on the extension of the Savant project [University of Cincinnati, 1999].

Savant allows the automatic translation of VHDL descriptions into C++ source code by using a two steps procedure. The first step converts the VHDL code into a intermediate functional and hierarchical representation called IIR. This representation defines a node, or sub-tree, for each VHDL statement (i.e., process, concurrent statement, sequential statement) and VHDL syntax element (i.e., variable, signal and constant). The C++ translation can be obtained by traversing the IIR tree. IIR features are exploited by the AMLETO extension to translate SystemC code into IIR, to inject faults into IIR, and, finally, to translate the modified IIR into SystemC code.

AMLETO extends SAVANT with the following modules depicted in Figure 5.5:

- **SCRAM SystemC Analyzer.** This module allows one to convert the SystemC description into the IIR format. SystemC up to version 1.2 can be directly converted into the IIR model, whilst new classes are added to IIR to describe SystemC 2.0 features;

- **IIR to SystemC translator.** Given an IIR structure this component can produce a SystemC description. This is used to generate the faulty SystemC description of the DUV, but it can be very usefull to reuse VHDL components in new SystemC-based projects as shown in Section 5.5;

- **Control module generator.** This module has been developed to allow the test of either each single SystemC module of a composite architecture or a set of them. Analyzing the modules' interconnections, it produces a new architecture description inserting a new SystemC module, called switch, over each interconnecting signal. This feature allows to propagate signals just to the target modules during simulation. The interface to switches is driven by the switch manager (see Section 5.4);

- **Fault list generator.** This object analyzes the IIR description to extract the fault list for the DUV according to the bit coverage fault model;

- **Fault injector.** This module gets either from the previous component or from file the fault list and produces an IIR description with injected faults;

- **TPG generator.** This component produces an architecture composed of: TPG, switch manager and fault detector. The generic TPG structure for testing a SystemC description is customized to the features of the DUV by this module (i.e., input/output port type, sequential/combinatorial circuit, etc.) and a test bench is automatically generated as explained in Section 5.4.1. A random and a genetic engine are available to generate a unique SystemC executable code to efficiently perform functional ATPG.

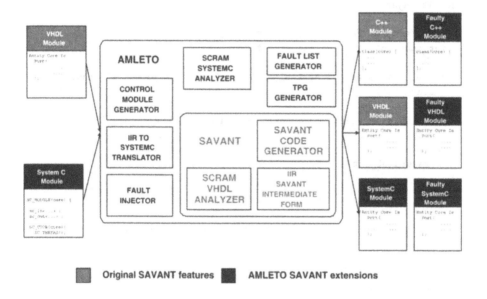

Original SAVANT features ■ AMLETO SAVANT extensions

Figure 5.5. AMLETO architecture

The SystemC description is perturbed by injecting all faults modeling by bit coverage. In this way perturbed SystemC code needs to be re-compiled and linked with the test bench and the ATPG engine just once. A control structure activates exactly one fault for each simulation run by means of saboteur functions, which replace the SystemC object (i.e., variable, signal, port, channel, ...) affected by the fault. These functions activate or de-activate the corresponding faults according to the value of an extra port, called fault_code, added to the faulty DUV. Each fault is identified by a numerical code which has to be written on the fault_code port by the ATPG. An injection fault rule has been defined for each basic statement, thus the fault injection is obtained analyzing each SystemC object and invoking the corresponding inject_fault function. Figure 5.6 shows the saboteur function for sc_bit operands. Parameters start_s0-1 and end_s0-1 in Figure 5.6, show the range for fault_code to activate the SA0-1 fault on the targeted statement. The SA0-1 masks allow to remove a specific fault if untestable, redundant or already detected. Figure 5.7 shows an example of fault-free and faulty SystemC descriptions. Moreover, it illustrates how the faults are recursively inserted in complex statements as an if-then-else statement. Figure 5.8 illustrates the fault injection architecture. It shows also an extra constant called delta. It has to be added to each component to allow the activation of a target fault even in descriptions with more instances of the same component.

```
sc_bit inject_fault (
  sc_bit object; int fault;
  int start_s0; int end_s0; sc_bit mask_s0;
  int start_s1; int end_s1; sc_bit mask_s1)
{
 if (fault >= start_s0 && fault <= end_s0 && mask_s0 == '1') then
   return('0');
 else
   if (fault >= start_s1 && fault <= end_s1 && mask_s1 == '1') then
     return('1');
   else
     return(object);
}
```

Figure 5.6. inject_fault SystemC function for sc_bit operands

```
if (data_in > rmax)
  rmax = data_in;

if (inject_fault(inject_fault(data_in,fault,
    1438,1445,"11111111",1446,1453,"11111111")
    > inject_fault(rmax, fault, ...),fault, ...))
  rmax = inject_fault(data_in,fault,
        1472,1479,"11111111",1480,1487,"11111111");
```

Figure 5.7. Fault-free and generated faulty SystemC code

5.4 High Level Validation of SystemC Design

The SystemC DUV code is linked to the DUV with injected faults and the ATPG tools to produce a unique executable program aiming at identifying untestable faults. Model checking is then used on these untestable faults to identify design errors. The architecture of the ATPG is described in the following subsection.

5.4.1   ATPG Architecture

Key part of the SystemC-based ATPG is the fault simulator. Being based on SystemC, it can be used to check the functional equivalence through all design phases (i.e., behavioral, RTL, gate). In fact, the concept of incremental test set can be applied to the presented ATPG. The inherited test set from the previous abstraction level is applied to detect all the level dependent faults of the current level description and is extended with new vectors able to detect specific level faults. This approach can reduce the testing time and it increases the fault coverage.

The SystemC simulation kernel allows one to trace the values of input/output signals of each module, whenever the proper SystemC instruction is added to the simulated SystemC code. However, during an ATPG session it is difficult to select a priori the set of signals to be traced. The main goal of the `switch` component is to increase the controllability and observability of each internal signal at simulation time. Moreover, a switched architecture can be tested either component by component or by grouping connected components. This latter method allows an incremental testing approach by testing sub-components of the whole architecture. Moreover, the insertion of switches allows to produce a single SystemC description, which is compiled only once. Each subarchitecture can be selected at run time by the `switch manager`. This feature increases the AMLETO efficiency avoiding the generation of descriptions for each sub architecture under test. Figure 5.9 shows a switch component.

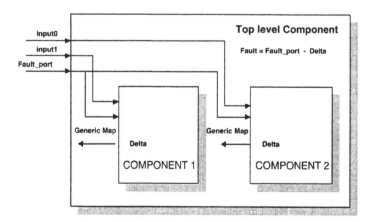

Figure 5.8.   Fault injection architecture

```
SC_MODULE(switch){
  sc_in <sc_bit> sigint;
  sc_in <sc_bit> exmode;
  sc_out <sc_bit> sigout;
  void filter();
  SC_CTOR(switch){
    SC_METHOD(filter);
    sensitive << sigin, exmode;
  }
}
void switch::filter{
  sc_bit var;
  ... ...
  switch(exmode){
    propagation: sigout = sigint;
    observation: var = sigint;
    control: sigout = var;
    ... ...
  }
}
```

Figure 5.9.   Switch component

The exmode signal sets the behavior of the switch: *propagate* or *isolate*. It is possible to test all the components and its internal connections by setting all the switches on the border of a sub-architecture to isolation mode and by enabling all its switches. If the exmode signal is set to propagation mode the value on sigin port is copied on the sigout port without modifying it. The value of the selected signal can be either read or write by using sigin and sigout signals.

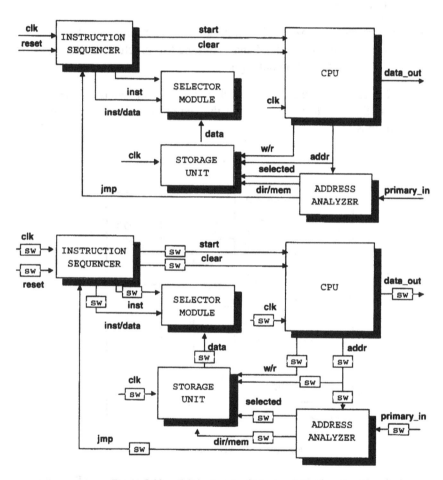

Figure 5.10. Original and switched architecture

Figure 5.10 shows an example of switched architecture. The configuration of all inserted switches is driven by the switch manager connected to the ATPG. The different sub architectures to be tested are read from a configuration file.

Figure 5.11 shows the architecture generated by the ATPG generator. It includes three main blocks:

140

- *ATPG engine*. This module produces the test vectors for the DUV. Its structure is customized on the DUV by analyzing the IIR structure. Input test vectors have to be of the same type of the DUV input ports. The module applies the same input vector to both faulty and fault-free descriptions, then, when the results are ready, the `fault detector` compares them. If the results are different for at least one of the observed output signals, the targeted fault has been detected, otherwise a new test vector must be applied;

- *Fault detector*. This component has to be configured for the DUV. Its configuration is obtained analyzing the IIR structure to check the data type of the observed output signals;

- *Switch manager*. This block is DUV independent. It selects the subarchitecture to test by setting the switches inside the two DUV descriptions.

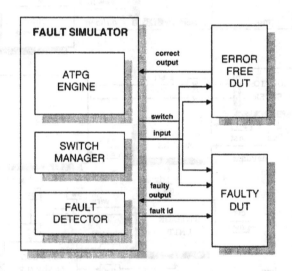

Figure 5.11. ATPG architecture

The structure of the ATPG allows the combination of many ATPG engines (i.e., genetics algorithms, BDD's, static code analysis, etc.) as described in the next section.

5.4.2 ATPG Engines

Random test pattern generation approaches allow to rapidly detect the most part of faults easy to detect. However, a large amount of time could be required to cover faults hard to detect. In the proposed SystemC based validation framework, the ATPG is aided by a genetic engine which is applied for extending randomly generated test patterns. Being SystemC based on C++, it easily allows the definition of an high accurate genetic algorithm:

The genetic algorithm developed for functional test generation has the following characteristics.

- **Gene definition and representation.** Genes are test vectors. Each one is GENE_SIZE long, where GENE_SIZE is the sum of primary input widths;

- **Population.** It is composed of 32 genes. Its size is constant during the whole genetic evolution. It evolves either for at most MAX_GENERATION generations or it stops if the GA reaches the full coverage;

- **Fitness function.** Several fitness functions have been tested before choosing for accuracy and performance:

$$\text{fitness}(X) = \frac{\text{faults detected by X}}{\text{remaining faults from previous generation}}$$

It requires the evaluation of the ratio between faults detected by the gene X and the remaining faults from previous generation. Both numbers are updated during the fault simulation, thus the fitness function can be evaluated efficiently;

- **Selection strategy.** The roulette selection scheme has been adopted. Experiments show its efficiency owed to its elitist factor. In fact, the top 5 genes from the previous generation take part of the gene set elaborated by genetic operators;

- **Crossover operator.** A GENE_SIZE-point crossover is applied. The probability of each crossover point is evaluated by the ranking derived from linkage learning theory. This solution allows to implicitly derive the points, which generate, on the average, better new individuals;

- **Mutation operator.** This operator slips gene bits. It allows one to detect faults from the same cluster. Its application probability increases when no coverage improvement is obtained during the last generations, otherwise it decreases to better explore fault clusters.

5.4.3 Model Checking

ATPG generally produces quite high fault coverage; however, some faults remain undetected, being redundant or hard to detect. Formal verification techniques, and in particular model checking, can be applied to better investigate the nature of these undetected faults.

Model checkers can work on a Boolean description of a DUV, which can be generated by using a synthesis tool. The analyzed SystemC must be thus synthesizable. Moreover, fault injection functions have been designed to be synthesizable in order to allow the application of a model checker on the faulty DUV.

Model checkers are used to verify the correctness of a system implementation according to the specifications formalized as properties written in temporal logic. For every rule, the model checker generates at least one sequence of stimuli that satisfies it (i.e., a witness) or that refutes it (i.e., a counter-example). However, if the complexity of the description is excessive for time and memory available on the computer running the tool, an inconclusive answer can be provided.

Based on model checking tools, we propose to write a set of formal properties aiming at characterizing the nature of the undetected faults. By proving these properties the model checker also generates a new set of test patterns, either witnesses or counterexamples, which can be provided to the fault simulator in order to increase the fault coverage. Thus undetected faults can be detected by simulating test patterns generated by proving properties. Two kinds of properties can be written:

- **Specific design dependent properties.** They are generated with respect to the specifications and the expected functionality of the system and they try to prove the two following characteristics:

  — *Safety*: to prove the correctness of the DUV by considering the fact that *the system must not evolve in a not accepted configuration*.

  — *Liveness*: to prove the reactivity of the DUV by considering the fact that *sooner or later the expected result, with respect to a particular input, should be obtained*.

  This kind of properties is suitable to prove the overall correctness of an implementation with respect to the corresponding specification, but an imprecise number of properties should be written in order to fulfill the purpose. On the other hand, we want just to focus the attention on some particular faults, those not detected by the ATPG. However, it is not an easy task to write properties that allow the model checker to generate test patterns for undetected faults, especially if the faults considered are injected into internal signals rather than into primary outputs. Moreover, a deep knowledge of the system is required to generate such kind of properties, thus automatic generation is difficult;

- **General output comparison properties** (OCP). They depend only on the primary outputs of the system and not on its functionality, thus no knowledge of the system behavior is required and automatic generation becomes feasible. The main idea is the same used for fault simulation. We assume the presence of a stuck at fault in the perturbed description of the system and we compare its outputs with the ones obtained by the fault-free system. The model checker should refute a simple property as one represented in Figure 5.12 in order to generate a test pattern identifying the analyzed fault. Figure 5.12 represents the main file written to prove the OCP property by using the Cadence SMV Model Checker [McMillan, 1993] on the synthesis reference example of SystemC Compiler [Synopsys, 2002a]. Note that the OCP property must be refuted in order to generate a counter-example that is able to detect the analyzed fault.

```
#include "VD_iq.smv"
MODULE main() {
  -- port declarations
  ...
  -- fault-free system instantiation
  ...
  -- faulty system instantiation
  ...
  -- error_port activates fault 2938
  error_port := 2938;
  -- formula to be proved
  -- po_i are primary outputs of the fault-free system
  -- po_i_e are primary outputs of the faulty system
  OCP: assert G((po_1=po_1_e)&...&(po_N=po_N_e));
}
```

Figure 5.12. Example of general output comparison property

Disregarding computational complexity, OCP properties are more suitable for our purpose, since they can be automatically generated. However, in some cases, symbolic model checking supplies inconclusive results.

In order to overcome this problem, a good choice is represented by *Bounded Model Checking*, a technique using a SAT solver to search for counterexamples of a bounded length. Finding counterexamples in this way is significantly faster than using symbolic model checking. We used the *zchaff* [Moskewicz et al., 2001] SAT solver joint with the SMV model checker since it allows to solve problems with more than one million variables and ten million clauses.

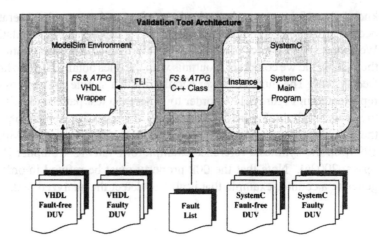

Figure 5.13. Validation tool architecture

5.5 Efficient Fault Simulation of a SystemC Design

The proposed validation environment is based on high-level fault simulation, which is a very time consuming activity. To improve its efficiency implications of the three kinds of simulations have been analyzed in this section:

- **Co-simulation.** A SystemC based implementation of the proposed validation tools allows to co-simulate heterogeneous designs composed of SystemC and other HDLs;

- **Simulation.** The validation architecture is linked to a pure SystemC model in the case that an automatic translation of other HDLs into SystemC code is provided;

- **Emulation.** The possibility to synthesize the faulty DUV on a hardware emulator allows its connection to the SystemC based ATPG and an impressive computation speed up.

5.5.1 Co-Simulation: the Validation Tool Architecture

Figure 5.13 shows the architecture of the proposed validation tool. The core of the application is represented by the *FS* and *ATPG* class. It is a SystemC class implementing the fault simulator and ATPG described in Section 5.4. It is simply instantiated by a SystemC program as well as by a VHDL wrapper.

SystemC simulation is performed by instantiating, in the main program, both the C++ class and the SystemC description of the design under validation. On the other hand, VHDL simulation can be acted by interfacing the C++ class and a VHDL simulator with a *Foreign Language Interface* library. A VHDL front end is included in order to test heterogeneous VHDL SystemC designs when proprietary VHDL core must be joint with new SystemC components.

5.5.2   Simulation: from HDL to SystemC

The majority of projects are not completely new, but they are based on predesigned components that are usually available at the RT level as VHDL (or Verilog) descriptions. The potential integration of such components must be carefully investigated by extensively simulating their interactions with the rest of the design. These simulation sessions can usually give enough information on the expected behaviors and performance. In this context the designer can tackle a new project by following two different strategies as described in Figure 5.14:

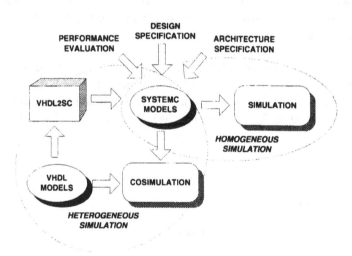

Figure 5.14.   Design strategies to mix VHDL and SystemC

- **Heterogeneous** modeling. A VHDL component selected from the library can be integrated with the SystemC description by using only a co-simulator [Valderrama et al., 1996, Soininen et al., 1995]. In this case, co-simulation implies an interface between the event driven VHDL simulation and the cycle-based SystemC simulation;

- **Homogeneous** modeling. The original VHDL descriptions are first translated into SystemC. The invariance of the original behavior can be guaranteed whenever event driven specifications can be described in cycle based terms. This is possible when all signals are updated synchronously with one or more clock signals. In this context the behavior of asynchronous signals becomes synchronous.

This section investigates the potentialities of the second approach. The manual re-modeling of VHDL components into SystemC modules is not feasible owing to the long time required to perform the translation, inherently prone to errors. Thus we have defined a set of rules aiming at automatically translate VHDL descriptions into SystemC modules with equivalent behavior under the assumption of cycle based simulation. In this way it is possible to obtain homogeneous representations, thus performing fault simulation instead of fault co-simulation sessions.

This translation has been implemented on top of the Savant environment as a further feature of the AMLETO environment (see Section 5.3.3). The SystemC code generator is based on the intermediate IIR format that is built from VHDL code. All identified translation rules have been implemented as C++ methods manipulating the IIR description.

The VHDL to SystemC translator Translation rules for all the VHDL constructs considered have been inserted in a database of rules and implemented in the vhdl2sc tool. Only partial examples are reported in the following for lack of space. For instance, Table 5.1 shows the rule to translate a fixed size integer.

Syntax	
VHDL	`integer range inf to sup`
SystemC	`sc_uint⌈< log₂(max(|sup|, |inf|) >⌉`

Example	
VHDL	`variable x :  integer range 0 to 15;`
SystemC	`sc_uint< 4 > x;`

Table 5.1. Translation of fixed size integer

In order to respect the semantic of the original VHDL description, some of the translation rules require a more detailed translation process than a pure syntax conversion. Figure 5.15 presents some interesting features related to the translation of VHDL models. The model considered describes a synchronous sequential machine. Owing to the `wait` statement the process has been translated by vhdl2sc into a SystemC SC_THREAD. The SystemC construct `wait_until()`, required to translate the VHDL `wait until`, can be inserted only in a SC_CTHREAD. Therefore we use the do-while cycle to achieve the same behavior. The VHDL variables have to be translated in C++ static variables in order to preserve their value for the next execution of the SystemC process.

```
ENTITY mouse IS                          SC_MODULE(mouse) {
  PORT(clk,xc,xd,rst:IN BIT;               sc_in<sc_bit> clk,xc,xd,rst;
       xp:OUT INTEGER );                   sc_out<int> xp;
END mouse;                                 sc_signal<int> pbs_state,x;
ARCHITECTURE bhv OF mouse IS               void pbs_machine();
  SIGNAL pbs_state,x:INTEGER;              SC_CTOR(mouse) {
BEGIN                                        SC_THREAD(pbs_machine);
  pbs_machine : PROCESS                      sensitive << clk;
    VARIABLE pbs_tok :INTEGER;             }
  BEGIN                                  };
    WAIT UNTIL clk'EVENT AND clk='1';    void mouse::pbs_machine() {
    pbs_tok:=0;                            static int pbs_tok;
    IF (reset='0') THEN}                   do { wait(); }
      ...                                  while (clk.event() && clk==1));
    END IF;                                pbs_tok=0;
  END PROCESS;                             if (reset==0) {
END bhv;                                     ...
                                           }
                                       }
```

Figure 5.15. Template of a synchronous sequential machine

Figure 5.16 shows the translation of an array variable. If there is no match between VHDL and SystemC data type, then the syntax can be different.

The VHDL case statement can be translated into the C++ switch, but the break instruction has to be inserted at the end of each branch. The others branch has to be declared as the C++ default branch.

A C++ function has to be defined to translate any not directly mappable VHDL arithmetic operator (e.g., the mod operator). There are some SystemC restrictions about vectors sc_bv and sc_lv, for instance, no arithmetic operators are defined for them. This translation problem can be solved by using the appropriate SystemC integer through cast operator (as shown in Figure 5.17) or support variables.

```
variable str : string(1 to 20);             char    str[20];
variable bv : bit_vector(1 to 8);           sc_bv<8> bv;
```

Figure 5.16. Different syntax for array type definition

```
VARIABLE xb:BOOLEAN;                          static bool xb;
VARIABLE xus,yus:UNSIGNED(7 DOWNTO 0);        static sc_uint<8> xus,yus;
xus:="11001100";                              sc_bv<8> xus_tmp,yus_tmp;
yus:="11001100";                              xus_tmp = "11001100";
xb:=(xus = yus);                              xus = xus_tmp;
                                              yus_tmp= "11001100";
                                              yus = yus_tmp;
                                              xb = xus == yus;
```

Figure 5.17. Support variable declarations

Both VHDL functions and procedures have to be translated into C++ methods. In the case of procedures, method parameters have to be passed by reference in order to model side effects of VHDL procedures into the calling process/procedure.

A namespace has to be defined to translate each VHDL package. It can include functions, procedures, variables, and constants. This approach allows the declaration of different objects with the same name in separated namespaces, thus behaving as a package. The translator is currently able to convert VHDL to SystemC by selecting the 1.2 or 2.0 version.

5.5.3   Emulation: from SystemC to Hardware

Homogeneous as well as heterogeneous models can be synthesized and then validate by using the same simulation and ATPG techniques presented in Section 5.4 running on high speed hardware emulators. In this way the defined *FS* and *ATPG* SystemC class is further extended to interact with the emulator.

Efficient hardware accelerated TPG techniques have been presented in recent works [Civera et al., 2001, Leveugle, 2000, Cheng et al., 1999]. The fault model adopted in [Civera et al., 2001] targets only memory elements and it reaches a fair average speed up, even if it can not be compared with the one obtained by the presented methodology. The design emulation presented in [Leveugle, 2000] is focused on 'in line' fault detection. Permanent single stuck at faults have been emulated in a FPGA based architecture in [Cheng et al., 1999].

In our approach the potentialities of an emulator to accelerate *FS* and *ATPG* aiming at identifying design errors are exploited. SystemC (and eventually VHDL) specifications are instrumented with faults and rapidly synthesized to be efficiently emulated (see Figure 5.18 and Figure 5.19).

The emulator is synchronized in different manners with *FS* and *ATPG* running on a host computer. These different methods induce different global performance. The identification of less expensive emulator configuration can have a relevant impact on the validation process.

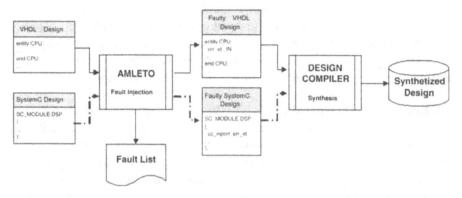

Figure 5.18. Heterogeneous system synthesis

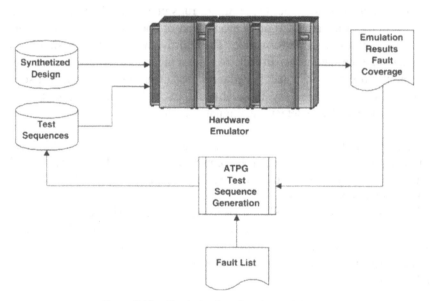

Figure 5.19. Emulation based testing architecture

Master/Slave Co-Emulation (cycle based). The workstation (master) running the ATPG downloads each single input test vector to the emulator (slave). The design outputs are read by the ATPG tool via emulator software interface.

The performance improvement that can be obtained depends on the complexity and the activity of the circuit. Running a software model, the simulation time is proportional to the number of events that have to be processed. On the other hand, the same functionality mapped on the emulator will run at a frequency which does not depend directly on the number of events. This configuration allows one to maintain a constantly updated vision of design configuration (signals and registers values). The ATPG tool samples the design primary outputs, after applying each test vector. This feature represents its main drawback, since the overhead due to the communication between the workstation is high.

Co-Emulation (Sequence Based). This implementation exploits the capability of loading memory contents on the emulator from a function call. A whole test sequence can be downloaded into the emulator's memory boards. If the capacity of the emulator's memory boards is higher than the space required to store a single test sequence, then the emulator's memory boards can be filled with several test sequences. This improvement allows to reduce the connection opening overhead per sequence, which produces a significant overall performance speed up.

The main advantage is that the emulator runs in a stand alone fashion since the system clock is generated on board instead of coming from the simulation side. This implies better performance. The improvement is the more significant the longer are the sequences. In fact the communication overhead between test generator and emulator is deeply reduced. To set up this configuration some additional work is required, synthesizable output checkers are downloaded into the emulator in order to detect the simulated fault.

Multiple-Fault Co-Emulation. Whenever, it is possible to download multiple copies of the design into the emulator, owing to the design size it is then possible to do parallel fault testing. This approach allows to reduce the overall validation phase cost.

The design fault list is split amongst the multiple design copies. Testing concurrently several design instances does not impact on the emulator performance and it can potentially further improve the testing performance. The best performance speed up is obtained if equivalent faults are simulated on the same DUT instance.

Performance Analysis. The required time to emulate a test sequence in the master/slave co-emulation mode is:

$$\mathtt{sequence_{etime}} = (t_{do} + t_{vd} + t_e) * \text{\# of vectors}$$

where t_{do} is the system download overhead (i.e. workstation-emulator communication setup), t_{vd} is the vector download time and t_e is the time required to emulate a single vector.

In the co-emulation mode the required time to emulate a test sequence is:

$$\mathtt{sequence_{etime}} = t_{do} + t_{sd} + (t_e * \text{\# of vectors})$$

where t_{sd} is the required time to download the whole sequence. Indeed, the co-emulation should be chosen for any architecture where:

$$t_{sd} < t_{vd} * \#ofvectors + t_{do} * (\# \text{ of vectors} - 1).$$

This constraint is valid for most of the available emulator systems. The master/slave co-emulation mode is more suitable at the initial stage of the design flow, when deep design behavior has to be verified.

5.6 Experimental Results

This section shows experimental results concerning all main aspects of the proposed validation methodology. Simulation and co-simulation is analyzed at first and then used as an infrastructure for testability analysis, which is further improved by using emulation.

5.6.1 VHDL to SystemC Translator Performance

The examined case study is composed of the AM2910 core (described in VHDL) and some surrounding blocks (designed in SystemC). The manual translation of the AM2910 in SystemC is a complex task. It requires an estimated effort of 30 hours, while its automatic translation in SystemC 2.0 requires a few milliseconds. Viable solutions for producing a simulatable model of the entire system are: the co-simulation of the heterogeneous model or the simulation of the automatically translated AM2910 description in SystemC.

	Real Time	User Time	System Time
VHDL only	1.5	1.2	2.1
VHDL+SystemC	10.3	7.5	5.3
SystemC only	1.0	1.0	1.0

Table 5.2. Simulation versus co-simulation, normalized times

VHDL simulation is performed by using a very efficient commercial VHDL simulator. The co-simulation approach is implemented by using a socket based interface between the executable SystemC description and the VHDL simulator. Simulation results are reported in Table 5.2 as normalized time with respect to the simulation of a homogeneous SystemC description of the entire system.

Simulation time required by the SystemC only description is sensibly lower (50%) than VHDL simulation. The co-simulation technique is the more complex task and it is sensibly slower than the SystemC based simulation, since it requires an event driven simulator to be interacted with a cycle based one. In conclusion, the homogeneous SystemC-based simulation shows better performance, with respect to the co-simulation approach, but also with respect to VHDL simulation. This last comparison is further investigated by examining different sets of VHDL constructs.

The first analysis concerns the presence of some function calls and parallel processes. We select three single process VHDL descriptions (ex1, ex2 and ex3) with a high number of function calls (respectively 97, 623 and 1283). SystemC descriptions are automatically generated and simulation times are measured for both VHDL and SystemC descriptions.

	ex1	ex2	ex3
VHDL	20%	57%	109%
SystemC	77%	177%	310%

Table 5.3. Percentage of delay introduced in the simulation by the function calls

	ex1	ex2	ex3
SystemC vs. VHDL	29.0	12.0	4.0

Table 5.4. Ratio between SystemC and VHDL simulation times with function calls

Table 5.3 shows the delay introduced by the function calls in the simulation of VHDL and SystemC descriptions stimulated with 100,000 input vectors. Results show that the function calls overhead in simulation time is higher in SystemC than in VHDL because of the minor complexity of SystemC processes regarding to VHDL ones. However, Table 5.4 shows that the simulation times of SystemC are clearly better than VHDL ones. Moreover, further experiments show that when the number of processes rises the advantage of SystemC over VHDL increases as well. Note that changing the C++ compilation optimizations and the way VHDL functions are translated in SystemC (e.g., static functions, 'in class' methods and 'in line' methods) impacts on SystemC simulation efficiency up to 44%.

The last set of experiments measures the impact of components instantiation in SystemC and VHDL. A structural VHDL description with a parametric number of components is selected and translated in SystemC. Simulation results are shown in Table 5.5. It is evident that the higher is the number of components the lower is the advantage of SystemC over VHDL. This observation can be

# Components	100	500	1000	5000	10000
VHDL	10.0	4.5	2.8	2.5	2.1
SystemC	1.0	1.0	1.0	1.0	1.0

Table 5.5. VHDL and SystemC simulation normalized times with respect to components instantiation

motivated by the minor complexity of SystemC processes regarding to VHDL ones.

In conclusion, the automatic translation of VHDL code into SystemC allows one to simulate a homogeneous SystemC model of an entire system even if some VHDL modules are reused in the design. SystemC simulation outperforms co-simulation and it is also more efficient than VHDL simulation particularly when the number of component instantiations and function calls per module is limited. This is usually the case for high-level models.

5.6.2 Functional Testability

The complex inverse quantizer of the MPEG encoder [Synopsys, 2002a] is considered as benchmark for verifying the main capabilities of the testability analysis techniques explained in Section 5.4.

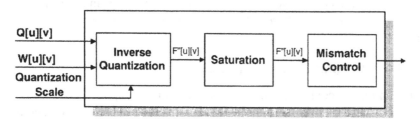

Figure 5.20. MPEG encoder architecture

Table 5.6 shows the fault coverage achieved and the ATPG characteristics related to the case study. The low fault coverage (69.9%) obtained shows a quite low testability of the analyzed design. The majority of the undetected faults are located within the quantization module mod.Q. By exploiting the controllability and observability capabilities, offered by the control switch architecture (Section 5.4), the inverse quantization module (mod.Q) is isolated from the saturation and the mismatch control modules (mod.S). These submodules are

Configuration	#Fault	Fault Cov.%	Time (sec.)	#Seq.	#Vect.
Whole design	3262	69.9	1863	68	49614
Quantization submodule	2478	78.4	3878	39	25011
Saturation & Mismatch submodules	784	83.2	792	19	11004

Table 5.6. Testability results

indenpendetly tested. As stated by the experimental results, the mod.S is not propagating most of the faults activated in the module mod.Q.

Moreover, a system refinement is considered from the testability point of view. The refinement includes the insertion of a hardware/software interface between modules mod.Q and mod.S. Different kinds of interfaces are simulated, and an interrupt-based parallel interface is considered. Some faults injected into the interface are undetected by the ATPG owing to controllability and observability problems highlighted by the control switch architecture. ATPG achieves 92.1% of fault coverage on the interface. The model checking approach, previously described, is applied to analyze the nature of these undetected faults, since they can be hard to test faults (corner cases) or redundant faults. Defined properties show that the 66.6% of undetected faults are actually redundant, whilst the remaining faults are simply hard to test. For these faults the model checker supplies a test pattern. A manual analysis of the redundant faults show that a part of the interface is useless, since data are not produced in time by module mod.Q to saturate the interface buffer. This is a typical over-specification design error.

5.6.3 Emulation Performance

The emulation configurations previously explained is applied to the AM2910 microcontroller used in the simulation/co-simulation experiments. The SystemC based ATPG runs on a Sun Ultra 60 with 1Gb of memory, whilst circuit emulation is performed on a leading emulator platform. The experimental results are presented in Table 5.7. Four different configurations are considered: two are purely software (VHDL and SystemC) and two are emulation based. The VHDL simulation is the slowest one and it has been used as reference for evaluating the performance speed-up of the other configurations. The emulation/simulation frequency is evaluated as follow:

$$F_{sim} = \frac{1}{T_{ps}}$$

where

$$N_{cyc} = \frac{T_s}{T_c} \qquad\qquad T_{ps} = \frac{T_{us}}{N_{cyc}}$$

N_{cyc} is the number of simulation cycles, T_s is the simulation time, T_c is the DUV clock period, T_{ps} is the required time to perform a simulation cycle and T_{us} the execution time on the workstation.

Configuration	Simulation/Emulation Freq. (KHz)	Speed up
VHDL	0.04	1
SystemC	2	50
Cycle-based	14	350
Sequence-based	98	2450

Table 5.7. Simulation times

The experimental results show the effectiveness of the adoption of a hardware emulator to accelerate the testing phases. In fact, the sequence based config-uration allows one to achieve a speed up of more than 2 orders of magnitude. The performance of the cycle based configuration is affected by the commu-nication bottleneck between the workstation and the emulator. This emulation configuration is suitable for accurate testing of small portion of the design, but the overall testing procedure of the whole design has to be performed by ap-plying the sequence based configuration. The emulator memory occupation for the considered benchmark allows to load several instances of it into the emulator boards. This configuration produces a further sub-linear performance increment.

5.7   Concluding Remarks

This chapter showed how SystemC 2.0 can be used as a complete verification environment for partial and complete SystemC designs. This is possible by defining a SystemC based fault model able to guide an ATPG and/or a model checker to identify hard to test and redundant faults. This testability analysis can be done at any abstraction level covered by SystemC 2.0. Moreover, a high computational efficiency has been achieved by exploiting the following points.

- The ATPG environment has been built by using the SystemC language in order to obtain a unique executable code composed of the DUV and the ATPG.

- A heterogeneous design composed of SystemC modules and HDL modules can be automatically translated into a homogeneous SystemC design to speed up the ATPG sessions avoiding co-simulation.

- A hardware emulator has been further used because the SystemC ATPG environment can be interfaced with the emulator itself, and the DUV with injected high-level faults is guaranteed to be synthesizable.

Further works are devoted to a more efficient SystemC based implementation of the ATPG and to study the integration of the proposed ATPG classes with the transactional level verification classes which are currently under development by the Verification SystemC Group.

Chapter 6

System Level Performance Estimation of Multi-Processing, Multi-Threading SoC Architectures for Networking Applications

Nuria Pazos,[1] Winthir Brunnbauer,[2] Jürgen Foag[1], Thomas Wild[1]

[1] *Institute for Integrated Circuits, TU-Munich, Germany*

[2] *Infineon, System Architecture, Data Comm., San Jose, CA, USA*

Abstract   The design of emerging networking architectures opens a multiple scenario of alternatives. If the design starts at a higher level of abstraction the process towards the selection of the optimal target architecture, as well as the partitioning of the functionalities, can be considerably accelerated. The present work introduces a novel system level performance estimation methodology based on SystemC. It will be shown that the rebuilding effort is considerably lower when applying the proposed methodology rather than building up a structural model of the target architecture at a lower level of abstraction. It makes feasible the exploration of several partitioning alternatives of a system specification onto a target architecture.

6.1   Introduction

High level estimation techniques are of great importance for design decisions, such as guiding the process of hardware–software partitioning. In this case a compromise between accuracy and computation time determines the feasibility of those estimation techniques.

The acceleration of the partitioning process, which allows the exploration of several partitioning alternatives, requires a fast performance estimation of the functionalities without losing accuracy. Time consumption of the whole design process could be drastically reduced if high level estimation methodologies would be performed before taking major design decisions.

W. Müller et al. (eds.),
SystemC: Methodologies and Applications, 157–190.

158

Figure 6.1 illustrates the two main different approaches that can be followed
when pursuing a first estimation of the performances achieved by a certain
partition alternative. On the one hand, building up a structural model of the
target architecture according to the selected partition of the functionalities.
It delivers a precise estimation, but the time and effort it takes to try a new
alternative is considerable. On the other hand, describing the functionalities in
terms of a Conditional Process Graph [Pop et al., 2000] and adding the relevant
information of the target architecture to the graph. This information comprises,
first, the performance of each function mapped to the selected processing unit
and, second, the characteristics of the communication architecture. In this
case the simulation time and effort towards a re-partitioning is less costly and
consequently more partition alternatives can be simulated and evaluated. After
choosing a partitioning that meets the performance constraints, the way towards
synthesis (back end) goes through a structural model of the target architecture.
This step is costly in terms of design, but fortunately it has to be done only once,
after selecting the best partition alternative that meets the constraints through
the loop described above.

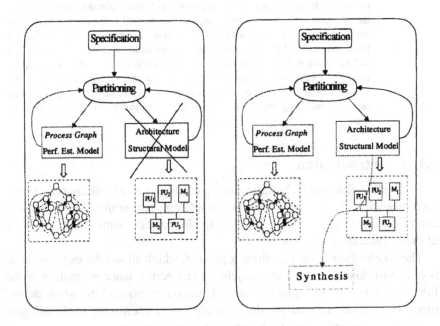

Figure 6.1.   Estimation of performance

The proposed methodology introduces the second approach. The building of an Annotated SystemC Conditional Synchronisation Graph (ASCSG) [Pazos et al., 2002], where each graph node is modelled in terms of a SystemC module, makes the mapping of the *computation nodes* onto the processing units and the corresponding addition of *communication nodes* much easier.

6.1.1   Application Scenario

The proposed method is oriented to support the design of multi-processing, multi-threading SoC architectures for networking applications. A packet processor [Husak, 2002] is a good example of such kind of architectures, where the independent packets delivered by the network have to be handled at high speed. That makes the parallel processing very suitable for this kind of applications.

Packet processing is ideal for an array of packet engines in the SoC. The packet engines are typically multi-threaded, in order to keep with the larger number of delivered packets. Many packet processors supplement the packet engines with coprocessors, implemented as fixed-function logic blocks. They speed up costly functions.

The abstract target architecture shown in Figure 6.2 contains diverse processing units (PU), memory units (M) and the required interface units to the external world. The processing units can be either low level RISC processors (PE) with multi-threading capability and no RTOS running on them, or hardware blocks (AC) accelerating certain functions. The memory units can be either global or local to specific blocks. The data communication between the different units takes place through a common bus (one or several dedicated ones), except the access to the local memories. When more than one master attempts to initiate an access to the common bus, an arbiter to police the requests is necessary.

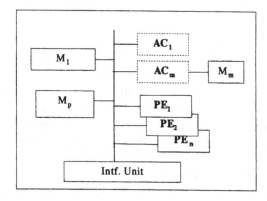

Figure 6.2.   Target architecture

The chapter is organised as follows: The first section introduces the motivation and the application scenario. Secondly, the Annotated SystemC Conditional Synchronisation Graph is introduced. It encloses the Functional, Architectural, and Communication models. The system performance estimation scheme is then illustrated by means of selective examples from the networking world. Later on, the chapter explains the steps towards the implementation and the integration process into the design flow. The performance results achieved by the proposed methodology are compared to the outcomes of a configurable cycle accurate hardware–software co-simulation platform for verification purposes. The next section shows and evaluates the results concerning computation time, re-building effort, and accuracy. The chapter ends with the conclusions.

6.2   State of the Art

The existing methods for automatic partitioning either ignore inter-component communication entirely or use simple models of communication to guide the partitioning–mapping step. The importance of integrating communication protocol selection with hardware–software partitioning are clearly demonstrated in [Knudsen and Madsen, 1998]. Furthermore, [Dey and Bommu, 1997] shows that the synchronisation overhead associated with inter-process communication can contribute significantly to the overall system performance.

The work dealing with the system performance analysis to help in the design of high performance communication architectures is very small in comparison with the work focusing on the partitioning–mapping step. The techniques that do consider the effects of the communication architecture can broadly be divided into the following categories:

Simulation Based Approaches.   These techniques are based on the simulation of the entire system. For this purpose different levels of abstraction in the modelling of the components and their communication are used. The level of abstraction for modelling the communication allows for a trade off between simulation time and accuracy. The approach presented in [Hines and Borriello, 1997] introduces a hardware–software co-simulator which provides substantial speed ups over traditional co-simulation methods by permitting dynamic changes in the level of detail when simulating communication channels between system components. In order to substantially improve the performance without sacrificing user access to detail, [Rowson and Sangiovanni-Vincentelli, 1997] proposes a simulator which separates communication from behaviour.

Static Performance Estimation Methods.   These techniques object the computationally prohibitive alternative of exhaustive simulation, arguing that an efficient exploration of the system design space necessitates fast and accurate performance estimation. They include models of the communication time between components of the system and often assume systems where the computations and communications can be statically scheduled [Gasteier and Glesner, 1999]. The communication time estimations used in these systems are either over optimistic, since they ignore dynamic effects such as wait time owed to bus contention, or are over pessimistic by assuming a worst case scenario for bus contention. This last case is addressed, for instance, in [Dey and Bommu, 1997], which issues a worst case performance analysis of a system described as a set of concurrent communicating processes.

Trace Based Performance Analysis Strategies. They are generally based on trace transformation techniques. They take as input a trace generated by an application process and generate, as output, a trace accepted by an architecture model and which contains the architecture level operations. Thus a trace transformation provides the mapping of application level communication primitive onto architecture level communication primitives[Lieverse et al., 2001]. Furthermore, there are some hybrid trace based performance analysis methodologies, as for example the one presented in [Lahiri et al., 1999], for driving the design of bus based SoC communication architectures, and in [Lahiri et al., 2001] for multi-channel communication architectures.

Analytical Performance Estimation Methods. The analytical methods are based on models of the tasks related to the application domain and a measure to characterise the performance of the target architecture. The goal of such methods is to identify interesting architectures quickly, which may then be subjected to a more detailed evaluation. In [Thiele et al., 2002] an interesting analytical method to explore the design space of packet processing architectures on the system level is presented.

The method proposed in the current work is a middle path between analytical and simulation based technique.

6.3   Methodology

In networking applications the processing of the packets crossing a network equipment depends on the values of certain fields within the packet/cell headers. This kind of system is suitable to be modelled using a Conditional Process Graph (CPG) [Pop et al., 2000] which is able to capture both data and control dependencies. Other possible graph based representations, such as the Control Data Flow Graph (CDFG) [Knudsen and Madsen, 1999], are not so adequate for describing networking functionalities. They are mainly thought for handling internal loops, which are found very seldom in networking applications.

The CPG proposed by Petru Eles describes each input functionality as a node. The nodes are linked in the defined order to achieve the complete system specification. Further on, by mapping each process to a given processing element and inserting additional processes, the so called *communication nodes*, on certain edges, a mapped process graph is built up. This graph contains the architecture characteristics of the final implementation of the functionalities onto the target architecture.

Based on the CPG and employing the support of SystemC, an Annotated SystemC Conditional Synchronisation Graph (ASCSG) is built. This description is then simulated in order to obtain an estimation of the performance values for the mapping–scheduling alternative under study.

For the modelling of the graph the industry standard systemlevel language SystemC has been chosen. It provides flexibility of modelling at various levels of abstraction and performs faster simulation. It also helps in modelling timing, reactivity, and concurrency, and in evaluating resource contentions. This makes this language very suitable for the implementation and further automation of the method–procedure.

The performance estimation methodology can be divided in two stages: Block level performance estimation and System level performance estimation. In the first stage the performance values for each sub-functionality mapped to each processing unit are calculated. This task is performed with the help of hardware and software processor models. These values are then stored in an internal library, which is accessed by the second phase. In the System level performance estimation stage the processing time of each sub-functionality, depending on the processing unit to which it is mapped, is annotated on it. Moreover, the system internal communication is considered, the conflicts when accessing the shared elements are solved, the transfer time through the communication medium is annotated, and the related queues are simulated. The current work concentrates itself on the second stage, the System level performance estimation. Nevertheless, the results of the Block level performance estimation are required before performing the second one.

This section is organised as follows: First, the models which build up the ASCSG are described. Then, based on this graph, the system performance estimation scheme is depicted and the related issues are explained in detail.

6.3.1 Annotated SystemC Conditional Synchronisation Graph

The ASCSG comprises three models, each one built on the top of the previous one: Functional, Architectural, and Communication model. The simulation of the last model delivers the pursued system performance values.

The definition of the former models and their main characteristics are presented in the following paragraphs.

Functional Model. This is the description of the system functionalities in form of a CPG. Each graph node, called a *computation node* (Cp), is modelled as a SystemC module. The functional validation of this model proves the correctness of the input specification.

The initial implementation of the graph, Figure 6.3, includes *computation nodes*. They represent the single processes into which the system functionality has been divided. A special case of *computation nodes* are the *condition nodes*. They represent the branching points in which, depending on computation results and/or external commands, the next branches to be taken are defined. The junction of such modules describes the complete system functionality in a structured manner.

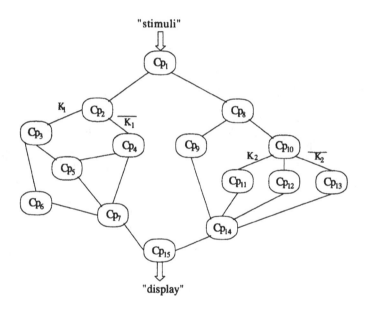

Figure 6.3. ASCSG functional model

Before starting the description of the Functional model, a decision concerning the granularity of the *computation nodes* has to be taken. A compromise between fine–coarse granularity (function level, command level or instruction level) has to be reached and the selected level has to be used further on in the definition and description of the *computation nodes*.

Architectural Model. Here takes place the mapping of each *computation node* (Cp) onto a processing unit (PU) which belongs to the previously defined target architecture (see Figure 6.4). Several instances of a certain node can also be mapped to different processing units. Furthermore, in the case of a multi-threading software implementation the mapping of the *computation nodes* to the related threads inside a processor is carried out. At this point the mapping of multiple instances of a node to different threads is possible as well. The result of this mapping is the annotation of each *computation node* with its execution time. An internal library provides the performance estimation of each block mapped to the selected processing unit. The parallel/sequential processing of nodes is hence determined, taking into account the restriction imposed by the mapping and the scheduling information.

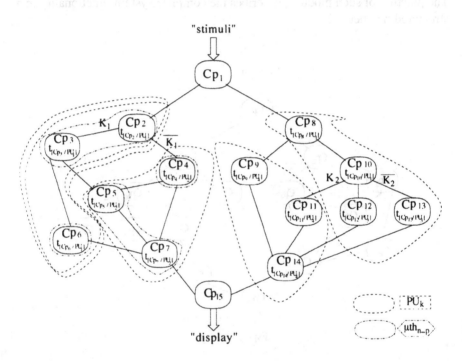

Figure 6.4. ASCSG architectural model

Communication Model. In this step additional nodes are introduced, the so called *communication nodes*, which define the communication cost between resources. They can either be external (Cme), between processing units, or internal (Cmi), between threads within a software processor. The Command Bus Arbiter (CBA) and Command Event Arbiter (CEA) for the inter-processing units and inter-threads communication, respectively, are added further on. The first one arbitrates the access to the shared communication media whereas the second one synchronises the threads within a software processor. These elements can be seen in Figure 6.5 where a section of a graph is depicted.

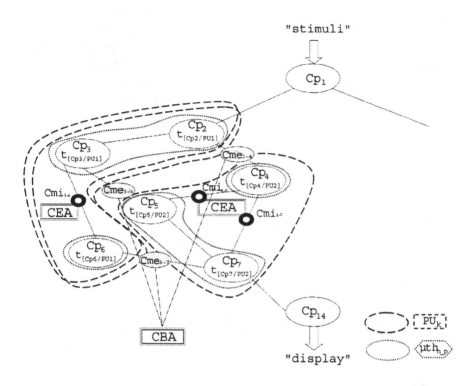

Figure 6.5. ASCSG communication model

The simulation of the Communication model delivers the performance characteristics required for the evaluation of a certain mapping–scheduling alternative.

6.3.2 System Performance Estimation Scheme

The estimation of the performance achieved by a certain mapping–scheduling alternative of a system specification onto a selected target architecture is carried out following the procedure depicted in Figure 6.6.

The single processes which make up the complete system functionality are selected from a library of functions. Such a library contains the description of each process in form of a SystemC module, which will be linked in the sequence previously defined in order to build up the Functional ASCSG. The test bench functions are additionally bound to the graph for feeding the model with stimuli and checking the results at the output.

The inputs required when pursuing the second model, the Architectural ASCSG, are the following. First of all, the structure and resources (kinds and numbers of processing units, memories and interface units) of the target architecture have to be provided. Furthermore, in case of software processors it has to be defined whether they support multi-threading and, in the affirmative case, the number of threads endorsed. Secondly, the information concerning the mapping of *computation nodes* onto the resources and threads has to be assigned. Lastly, the pre-defined scheduling of the *computation nodes* is also taken as input in order to define the overall flow of functions.

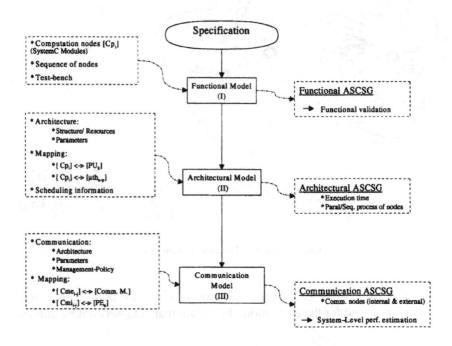

Figure 6.6. System performance estimation scheme

Once the mapping is fixed the related execution times of each *computation node*, depending on which processing unit is mapped, are fetched from an external library and annotated into the corresponding nodes. Moreover, the mapping and scheduling information define the parallel/sequential processing of nodes.

The last model, the Communication ASCSG, accepts as inputs the structure of the communication architecture, the related parameters for each communication medium (bandwidth and frequency) and the management policy implemented when accessing a shared communication medium and when sharing a processing unit by several threads. Furthermore, the mapping of the *communication nodes*, the external ones onto the communication media and the internal ones onto the multi-threading software processors, is performed.

During the generation of the models mentioned above different issues have to be considered. The definition of each issue and the proposed approaches to solve them are further presented.

Sequence of Nodes.   The *computation nodes* in the Functional model are triggered by events, either input data changes or positive/negative edges in case of boolean inputs, as can be seen in the Figure 6.7. The sensitivity list of each process inside each *computation node* defines the signals or ports, which trigger the corresponding process.

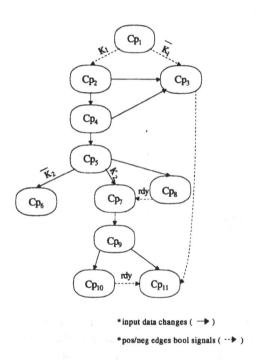

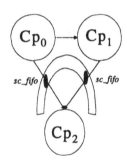

Figure 6.8. Convergence of paths

* input data changes (→)

* pos/neg edges bool signals (--▶)

Figure 6.7.   Sequence of nodes

Convergence of Paths.   Within a process graph it can occur that the outputs of two or more *computation nodes* sourcing from different branches which are not exclusive to each other meet together. At this point it can not be assured that the upper nodes deliver the outcome with the same latency. Therefore the meeting node has to wait until all required inputs are available. In order to avoid the loss of intermediate results, a *fifo channel* (SystemC primitive channel sc_fifo), as depicted in Figure 6.8, has to bind each upper node with the meeting one.

Functional Validation.   The Functional model has to be functionally validated in order to check the correctness of the specification. For this purpose, a test bench is built around the model, which can be re-used in further models later on. The test bench, Figure 6.9, comprises a module to generate the stimuli and a module to visualise the results.

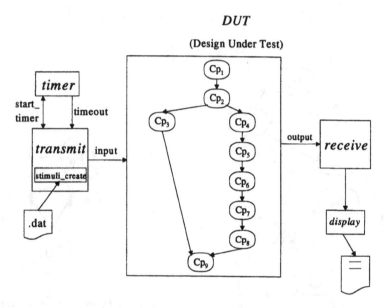

Figure 6.9.   Functional validation

Flexible Mapping.   This technique allows the exploration of different mapping–scheduling alternatives without requiring a re-building of the structure. It is possible owing to a flexible mapping of the *computation nodes* onto the hardware units and onto processors and threads, Figure 6.10, which belong to the target architecture. An internal library provides the performance estimation of each node mapped to the selected processing unit. The annotation of each *computation node* with the respective execution time (x), depending on the processing unit it is mapped, is carried out using the SystemC statement wait(x,SC_NS). The usage of timing control statements is only possible in SC_THREAD or SC_CTHREAD processes. Hence, the processes involved have to be defined as SC_CTHREAD, if they are sensitive to an edge of a clock, or as SC_THREAD in the general case. At this stage special attention has to be paid because multiple interacting synchronous processes must have at least one timing control statement in every path.

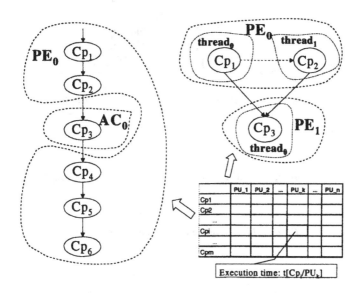

Figure 6.10. Mapping onto PE/thread–AC

Multiple Instances of *Computation Nodes*.   In the case of multiple processing units or threads performing the same functionality, as shown in Figure 6.11, the concerned *computation nodes* are defined only once and instantiated several times. In SystemC this can be performed defining a pointer to the related *computation node* (Cp_i *) and making several instances of the node in the main file, which can be customised at elaboration time by passing constructor arguments.

Detection of Shared *Computation Nodes*.   The *computation nodes* triggered by diverse sources running in parallel have to be detected and a mechanism to store the requests and process them in a determined order, for instance giving different priorities to the source nodes, has to be provided. Such a mechanism is referred to as a *shared element*. Moreover, an *access node* has to be interleaved between the shared *computation node* and the node which attempts to access it, for writing or reading. Such *access nodes* are bound to the *shared element* mechanism and keep the requests until the *shared element* grants the access to the shared *computation node*, as shown in the Figure 6.12.

The shared *computation nodes* can be modelled as a SystemC sc_module or as SystemC hierarchical channel in case they implement an interface (e.g. a shared memory). In the last case the corresponding *access node* also

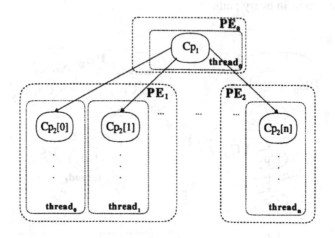

Figure 6.11. Multiple instances of *comp. nodes*

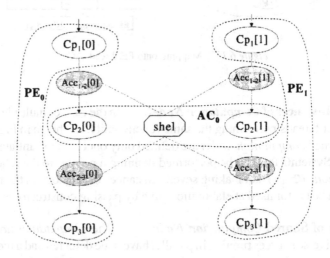

Figure 6.12. Shared *comp. node*

has to implement the interface functions in order to maintain the original graph unchangeable.

Parallel/sequential Procedure of Threads within a Multi-Threading Software Processor. In a multi-threading software processor there is a fixed number of threads running in parallel, but sharing the processing unit. Once an active thread performs a context switch, for instance after sending a request to an external unit, it has to wait until the request is served. The thread then suspends its execution and surrenders the access to the processing unit to another thread. Once the request has been performed the external unit signalises it to the corresponding thread, which is again ready for being awake. Anyhow, it has to wait until the processing unit becomes free and the Context Event Arbiter (CEA) grants it the access.

In order to be able to implement this characteristic the context switch events have to be signalised to the Context Event Arbiter, whose modelling will be explained within the specification of the Communication model.

Scheduling. The scheduling information, i.e., the determination of the sequence in which the operations are carried out, is also required at this point. As default an ASAP scheduling procedure is supported, in which each operation is performed as soon as the previous one has been completed. The only restrictions imposed are the ones concerning the access to shared *computation nodes*. Each initiator owns a priority number for accessing a shared *computation node*, which will be arbitrated in a way determined by the management policy implemented by the corresponding *shared element* mechanism.

Insertion of External *Communication Nodes*. Each time two consecutive nodes are mapped to different processing units an external *communication node* (Cme) must be inserted in the graph, Figures 6.13 and 6.14. In the case in which between two consecutive nodes more than one data transfer takes place, the corresponding number of external *communication nodes* has to be inserted. For the case of a bus based communication architecture the following aspects have to be considered during a data transmission:

- The initiator processing unit first signals the Command Bus Arbiter (CBA) its intention through the external *communication node*. It will grant the access to the target communication medium;

- A management policy for the CBA of each communication medium has to be selected. A default priority based policy is implemented.

Insertion of Internal *Communication Nodes*. When two consecutive nodes are mapped to different threads, an internal *communication node* (Cmi) must be added, as shown in Figure 6.14. During an inter-thread communication the next has to be contemplated:

- The context/s ready to run signalise the Context Event Arbiter (CEA) their status (waiting for a context switch event) through an internal *communication node*;

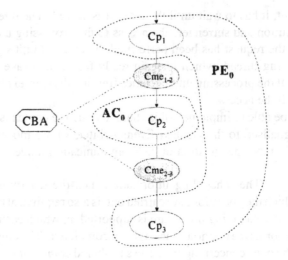

Figure 6.13. External *comm. nodes*

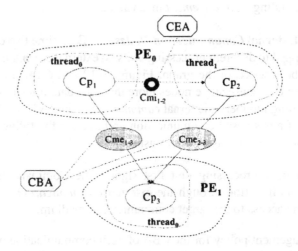

Figure 6.14. Ext./int. *comm. nodes*

- Once the active thread performs a context switch it is signalised to the CEA through an internal *communication node*;

- A management policy for the CEA of each software processor, in order to decide which is the next thread to run, has to be selected.

Model of the Shared Communication Media. The external *communication nodes* added to the graph solve the access to the shared communication media. Therefore each one accesses the Command Bus Arbiter of the communication medium to which it is mapped.

The target architecture comprises masters (or transmission initiators) and slaves (or transmission receivers). When several masters are able to access the same slave the last is referred to as *shared element*. In a bus based SoC architecture the access to the *shared elements* takes place through a common bus. Inside each shared communication medium a *command fifo* is implemented, where the requests stay until they can be served.

Tracking and Monitoring.　　From the simulation of the Communication ASCSG the following values are extracted, whose evaluation determines the performance estimation of the system: Processing time of each processing unit; delay accessing each communication medium; arbitration delays; delay accessing each *shared element* and its load; delay accessing the process-unit inside each software processor by each thread and context switch delay.

6.4　　Implementation Procedure

The construction of the models related to the mapping–scheduling alternative under study requires some input information. This concerns, for instance, the functionalities to be validated, the resources and structure of the target architecture, the selected mapping and scheduling alternative, and the management policies when accessing shared functions or shared communication media.

The exploration of several mapping alternatives, and even the inspection of different target architectures, is feasible, provided that the user iteration can be automatised. By the usage of configuration files the method process can be fed with the required input parameters. A parser will further analyse the syntax of the corresponding configuration file, and lastly a generator will create the related executable model.

First of all, this section shows the implementation procedure towards the creation of the Functional, Architectural and Communication model, and, later on, the data structures and configuration files used when pursuing the automation of the method process are presented.

6.4.1　　Implementation of the ASCSG

The procedure towards the construction of the Annotated SystemC Conditional Synchronisation Graph is depicted in Figure 6.15. It consists of three steps which are shared by each model, the Functional, the Architectural, and the Communication model. The first column shows the required input parameters, files, and templates for each model. With this information a generator creates the corresponding configuration files, each on top of the previous one. These configuration files are then parsed in order to generate the correspond-

174

ing executable file, which is finally executed in order to perform the functional validation of the specification and the extraction of the performance values for the selected alternative.

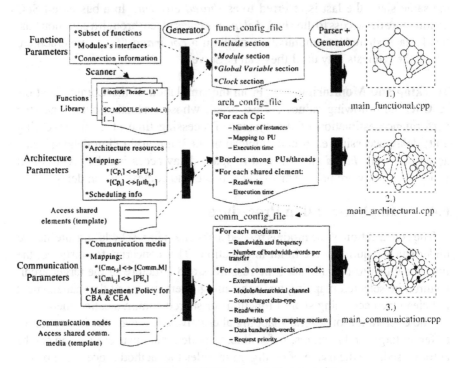

Figure 6.15. ASCSG implementation procedure

Creation of the Functional model. The system functionalities are described in terms of a Conditional Process Graph. The granularity of the graph is assumed to be at the level of function calls, which represent certain protocol specific functions within the desired packet processing application. When choosing the granules the memory organisation has to be taken into account, because internal transfers are only possible when a node has finished execution. Each graph node, referred to as *computation node*, is described in terms of a single SystemC module and stored in a library of functions.

When building up the Functional model for the selected application, the subset of required functions is defined. After that a scanner looks inside the library of functions to extract the interface information for each SystemC module. Subsequently an user interface allows the designer to introduce the connections between modules. A port type of and data type of check is thereby performed. Taking into account the previous information, a functional configuration file is generated, which further on is parsed for the creation of the functional executable file.

Creation of the Architectural Model.   The structure and the resources of the target architecture have to be provided when pursuing the Architectural model. Therefore a list is defined which contains the available processing units (PU) (either low level RISC processors (PE) with multi-threading capability and no RTOS running on them, or hardware blocks (AC) accelerating certain functions), memory units (M) and interface units to the external world. Furthermore, for each software processor the number of supported threads has to be specified.

The processing time of each *computation node* mapped to the diverse processing units is predetermined. These execution time values are stored in a library, Table 6.1, which is accessible by the mapping process. The boxes with an X mean that the related mapping is not feasible.

Table 6.1.   Mapping processing times

	PU1	PU2	...	PUk	...	PUn
Cp_1	$t[Cp_1/PU_1]$	$t[Cp_1/PU_2]$	—	$t[Cp_1/PU_k]$	—	$t[Cp_1/PU_n]$
...	—	—	—	—	—	—
Cp_i	$t[Cp_i/PU_1]$	X	—	$t[Cp_i/PU_k]$	—	$t[Cp_i/PU_n]$
...	—	—	—	—	—	—
Cp_m	$t[Cp_m/PU_1]$	$t[Cp_m/PU_2]$	—	X	—	$t[Cp_m/PU_n]$

The mapping process performs the assignment of *computation nodes* onto the available processing units and, in the case of the implementation of software, threads within a processor. Parallel processing units carrying out the same function arises frequently in networking applications. Therefore the *computation nodes* which make up the initial graph can be instantiated more than once and mapped to different processing units and threads.

Summarising, the information contained in the architectural configuration file for each *computation node* is: Number of instances; mapping of each instance (processing unit number and priority); and, depending on it, its execution time, which is extracted from the Mapping Processing Times library, Table 6.1.

After the mapping has been performed the borders between processing units are detected. Between two consecutive nodes which belong to different processing units and are not shared by other nodes, a *fifo channel* is introduced. It avoids the loss of information between such nodes. The case of *computation nodes* triggered by diverse sources running in parallel is solved by providing a mechanism, which police the requests coming from different sources attempting to access the shared *computation node* at the same time or while it is busy.

In case of a simultaneous access attempt, the policing mechanism takes into account the priority of the initiators to decide which source *computation node* accesses the shared one first.

A template for the policing mechanism, referred to as *shared element*, is provided, shel<SHEL_EXEC>. For its customisation the processing time of the shared *computation node* is required. Moreover, an *access node* is interleaved between the initiator and the shared *computation node*. It is responsible for requesting the read/write access to the *shared element* and wait until it grants access. A template for it is also available, acc<T, R_W>, which is customised providing the data type of the access initiator, whether it is a read or write access and the priority of the request. This last parameter is passed as a constructor argument to the module. In the case in which either the transmission initiator or the receiver is implemented as a hierarchical channel (which implements the ch_if<T> interface) the template acc_ch<T, R_W> might be used instead.

```
template <int SHEL_EXEC>
  class shel: public shel_if, public sc_module {...};
template <class T, bool R_W>
  class acc: public sc_module {...};
template <class T, bool R_W>
  class acc_ch: public sc_module {...};
```

Figure 6.16 depicts the access for writing and reading of two instances of two *computation nodes* mapped to two different software processors (PE), to a shared one mapped to an accelerator (AC). For each data to be transfered an *access node* (Acc) has to be included. It accesses the corresponding *shared element* (shel) through a port which is able to call the interface functions implemented in the *shared element*.

Creation of the Communication Model. The last model, the Communication model, includes the characteristics of the two previous ones and introduces the communication architecture. A list of the available communication media is thus defined for the external communication as well as the available threads within each software processor for the inter-thread synchronisation.

Each time a transfer through a certain communication medium takes place, an external *communication node* (Cme) is inserted, as shown in Figure 6.16. Such a node provides the access of a processing unit to the selected communication medium. It also calculates the transit duration for a certain transfer, taking into account the data to be transmitted and the bandwidth and frequency of the communication medium. Each external *communication node* has to be mapped to the medium where the transfer takes place. A template for it is provided,

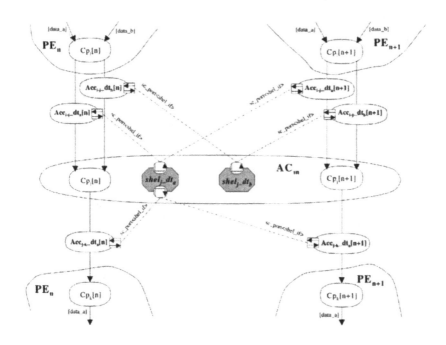

Figure 6.16. Access to a shared *computation node* through a shared communication medium

cme<T, M_BW, M_LW_T, R_W>, whose customisation requires the following parameters: Source/target data type; request type (read or write); bandwidth of the mapping medium; number of words of bandwidth to be transmitted; and the priority of the request. This last parameter is passed as a constructor argument to the node. In the case in which either the transmission initiator or the receiver is implemented as a hierarchical channel (which implements the ch_if<T> interface) the template cme_ch<T, M_BW, M_LW_T, R_W> might be used instead, as seen in Figure 6.17.

> template <class T, int M_BW, int M_LW_T, bool R_W>
> class cme: public sc_module {...};
> template <class T, int M_BW, int M_LW_T, bool R_W>
> class cme_ch: public channel_if<T>, public sc_module {...};

Furthermore, each communication medium implements a management policy, the Command Bus Arbiter (CBA), in order to select the first request to be served. It is accessed by the related external *communication nodes* through a port which is able to call the interface functions implemented in the Command Bus Arbiter. A template for it is also available, cba<M_BW, M_LW_T, M_FREQ>. For each communication medium a CBA has to be included, which is accessed by the corresponding external *communication nodes*.

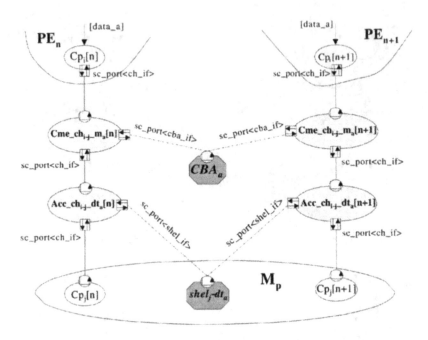

Figure 6.17. Access to a shared communication medium from/to a hierarchical channel

The required parameters for its customisation are the bandwidth, frequency and the number of words of bandwidth to be transmitted at once.

$$\text{template } <\text{int M\_BW, int M\_LW\_T, int M\_FREQ}>$$
$$\text{class cba: public cba\_if, public sc\_module } \{...\};$$

Analogously, each time a synchronisation between threads has to be modelled, an internal *communication node* (Cmi) is inserted. A template is provided for this purpose, cmi<T, Ev_Sw>, whose customisation requires only the source/target data type, whether the node waits for an event or signalises a context switch and the thread number which initiates the request. This last parameter is passed as a constructor argument to the node. In the case in which either the transmission initiator or the receiver is implemented as a hierarchical channel (which implements the ch_if<T> interface) the template cmi_ch<T, Ev_Sw> might be used instead.

$$\text{template } <\text{class T, bool Ev\_Sw}>$$
$$\text{class cmi: public sc\_module } \{...\};$$
$$\text{template } <\text{class T, bool Ev\_Sw}>$$
$$\text{class cmi\_ch: public channel\_if}<T>, \text{public sc\_module } \{...\};$$

The synchronisation between threads within a software processor is policed by the Context Event Arbiter (CEA), which is accessed by the internal *communication nodes*. It also implements a management policy in order to decide which is the next thread within the processor to run. Each internal *communication node* has to be mapped to the software processor, whose threads synchronises. For each multi-threading processor a CEA is instantiated, whose customisation requires the number of supported threads. A template cea<N_UTH> is provided for this purpose.

> template <int N_UTH>
> class cea: public cea_if, public sc_module {...};

6.4.2   Automation Approach

Once the input functional specification and the target architecture, the building blocks, and the communication structure, have been defined the exploration of several mapping–scheduling alternatives requires only the information concerning the mapping of the *computation nodes* onto the processing units, the mapping of the external *communication nodes* onto the communication media and the scheduling information. With these parameters the implementation procedure presented in the previous section is automated and the Communication ASCSG for every mapping–scheduling alternative extracted. The evaluation and comparison of such output graphs give the designer the required references for choosing the most adequate alternative.

Data Structure for the Implementation.   A common design approach is followed with the data structures shown in Figures 6.18, 6.19 and 6.20, which are further annotated in subsequent steps. These sample data structures are intended to clarify the semantics of the automation.

For each *computation node* a data structure with the following fields, as shown in Figure 6.18, is built up: Name [CP_NAME]; identification [ID]; whether it is modelled as a hierarchical channel or as a module [HC_MOD]; and number of instances [NUM_INST]. Furthermore, the data structure comprises a pointer to the contained ports, port_list, and a second one to the mapping information of each instance, mapp_list.

In the same way the required information for each port is: Name [PT_NAME]; data type [DT]; and whether it is a SystemC/C++ built in or an user defined data type [UDT]; port mode [PM]; whether it is bound [BN]; and the name of the signal attached to it [SIG_N]. Moreover, also a pointer, opp-1, to the attached *computation node* [CP] and the related port [PT] is added.

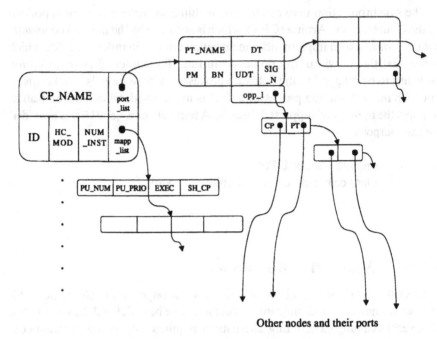

Figure 6.18. Computation node data structure

The mapping process associates to each instance the parameters concerning the processing unit number [PU_NUM], the priority of the processing unit [PU_PRIO], its execution time [EXEC], and whether it is a shared node [SH_CP].

As has been seen in the previous section, the detection of a *computation node* triggered by multiple sources implies the insertion of an *access node* in between and a policing mechanism, referred to as *shared element*. The data structures for these two elements are shown in Figure 6.19.

Each *access node* contains its name [ACC_NAME], identification [ID], data type to pass across [DT], and whether it attempts a read or a write request to/from the shared *computation node* [R_W]. A pointer to the list of its ports, port_list, gives access to the information related to each port: Name [PT_NAME]; data type [DT]; and port mode [PT_MOD]. A further pointer, opp-1, binds each port either to the concerning *computation node* [CP] and the related port [PT], or to the policing mechanism, the *shared element* [SHEL].

The information required for each *shared element* in order to further customise the related template is: Name [SHEL_NAME]; identification [ID]; and the execution time of the shared *computation node* [SH_CP_EXEC].

The last step towards the creation of the communication ASCSG comprises the insertion of internal and external *communication nodes* for the synchronisation between threads and the access to the communication media, respectively.

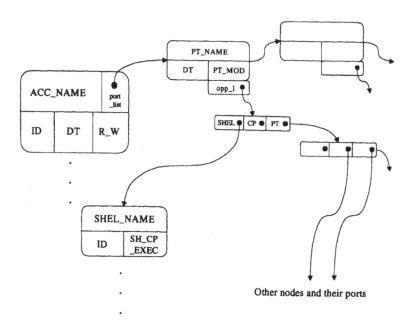

Figure 6.19.   Access node / shared element data structure

The customisation of such additional nodes requires some parameters, which are further annotated in their data structure as shown in Figure 6.20.

Each *communication node* data structure comprises its name [CM_NAME], identification [ID], type (internal or external) [TYP], whether it is modelled as a hierarchical channel or as a module [HC_MOD], whether it attempts a read or a write request [R_W] in case of an external communication node or whether it waits for an event or signalises a context switch in case of an internal one, bandwidth of the communication medium[1] [M_BW], number of bandwidth words to be transmitted at once through the medium[1] [BW_WD_NUM], and priority of the initiator [PRIO]. A pointer to the list of its ports, port_list, gives access to the information related to each port: Name [PT_NAME], data type [DT] and port mode [PT_MOD]. A further pointer, opp-1, binds each port either to the concerning *computation node* [CP] and the related port [PT], or to the policing mechanism, the *shared element* [SHEL], or to the arbitration mechanism [ARB], the Context Event Arbiter and the Command Bus Arbiter for the internal and external *communication nodes* respectively.

---

[1] Only required in case of an external *communication node*

182

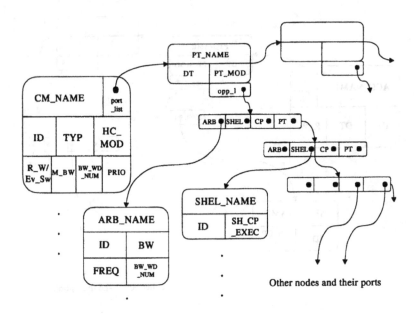

Figure 6.20.   Communication node / arbiter data structure

The information required for each arbitration mechanism in order to further customise the related template is: Name [ARB_NAME]; identification [ID]; bandwidth[1] [BW]; and frequency [FREQ][1] of the communication medium and number of bandwidth words to be transmitted at once through the medium [BW_WD_NUM].

Configuration Files.   The Functional model is created using a subset of the library of functions and the explicity interconnection information of their ports, which is completely specified in the functional configuration file.

Since the task of specifying the interconnects is bit consumed, an option of interactively creating the initial configuration file is implemented to which the subset of functions is the input. The input files are scanned for the port definition, types, names, and only the feasible links are displayed from which the user could select the appropriate one. The configuration file catering to the Functional model could also be tuned manually, thus saving time for creating it, right again from scratch. This configuration file is then parsed and fed to the data structure defined above in Figure 6.18 (except the elements of the mapping list whose are defined after the mapping process). Finally, these linked lists are simply traversed and processed through iteratively in order to generate the simulatable Functional model (*main_functional.cpp*).

Specifically, the simulatable Functional model is a SystemC main file which is divided into the following sections: Include files; channel declaration; module instantiation; and trace file section. Thus the functional configuration file consists of the following sections:

- Include section: contains the name of the header files to be included;

- Module section: comprises the name of the modules along with their parameters and the list of all their ports and port's details. Moreover, it contains the list of modules and ports to which each of the own ports are connected;

- Global Variable section: includes more general information as for example the desired name of the output file, the trace file name and its type, i.e., vcd or wif;

- Clock section: specifies the details for every clock and their parameters;

- Channel section: defines the links between ports, either through SystemC hardware signals or through SystemC primitive channels.

The steps towards the Architectural model imply the mapping of every instance of each *computation node* to the selected functional unit of the target architecture. This is done by making the *computation node* hold more information inside its data structure. This information includes, for each instance, the processing unit number to which it is mapped, its priority, and the corresponding execution time, which is picked up from the Mapping Processing Times library.

With these parameters a process detects whether shared *computation nodes* exist and, in the affirmative case, the corresponding flag inside its data structure is set. Furthermore, a list of required *access nodes* and another of the corresponding *shared elements* are created. The required information for their customisation is extracted from both the access initiator and the shared *computation node*.

In practice, the architectural configuration file extends the previously functional configuration file. The Module section additionally comprises the mapping information of each *computation node* instance and the lists of *access nodes* and *shared elements* along with their details. This architectural configuration file is then parsed and fed to the data structures defined in Figures 6.18 and 6.19. Finally, these linked lists are simply traversed and processed through iteratively in order to generate the simulatable Architectural model (*main_architectural.cpp*).

The last step pursuing the Communication model carries out the insertion of internal and external *communication nodes* and the corresponding arbitration mechanisms to access the shared processors and media respectively. For these additional elements two new linked lists are created. The mapping of the internal and external transmissions to the target software processors and communication media, respectively, gives the input information required by their customisation.

The communication configuration file extends the Module section with these two new linked lists of *communication nodes* and arbitration mechanisms. This configuration file is again parsed and fed to the data structures defined in Figures 6.18, 6.19 and 6.20. Once again and lastly, these linked lists are simply traversed and processed through iteratively in order to generate the simulatable Communication model (*main_communication.cpp*).

6.5   Methodology Verification

For the performance and accuracy verification of the proposed methodology, a cycle accurate model of the system to be evaluated is required. The analyse of the performance achieved are compared with the estimation provided by the previous methodology. The same procedure is repeated for diverse architecture configurations, trying to cover most of the possible scenarios. A flexible hardware–software co-simulation platform substantially facilitates this step. It consists of several modules which simulate hardware specific blocks, software running into embedded RISC processors and their interactions with memory blocks and interfaces with the external world.

The significance of the current work lies on the usage of the same language, SystemC, as specification language for the system level performance estimation methodology and for the modelling of the platform. This issue makes the comparison of results of both models easier and reliable.

As base for the development of the co-simulation platform, the target architecture depicted in Figure 6.2 has been chosen. A Cycle accurate Co-simulation Model is performed, using for it an Instruction Set Simulator (ISS) of the multithreading software processor. It provides accurate cycle counts required by the cycle accurate performance analysis. A cycle accurate model of the hardware blocks is also needed at this point. The functions of the system specification are translated into assembler code which is read by the ISS.

6.5.1   Flexible Hardware–Software Co-simulation Platform

A flexible hardware–software bus based co-simulation platform is able to simulate and verify complex architectures which involve embedded software processors and hardware blocks. The number of such internal processing units is thus configurable in order to decrease design times and find the best target architecture–partitioning–scheduling alternative which meets the predefined constraints.

In this case the data/command transmission between resources is restricted to a bus based communication, except the point to point link to access the local memories. A more general platform should also support a configurable communication structure, but it is out of the scope of the current work.

Figure 6.21 depicts the implemented co-simulation platform. As introduced above, the number of multi-threading software processors (PE) and hardware blocks (AC) acting as accelerators of certain costly functions is configurable. Each unit has an identification number which is further on used as priority number when solving concurrent accesses to a *shared element* or to a shared communication medium.

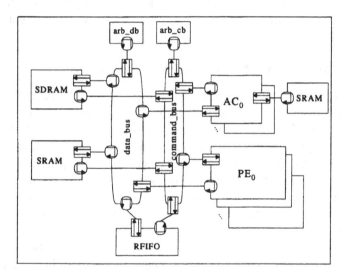

Figure 6.21. Bus based co-simulation platform

The embedded software processors work in parallel controlling the whole processing of the packets entering the system. First of all, they pick up the packets from the receive fifo (RFIFO), where the packets are stored in the arrival sequence. Secondly, the processors start performing the sequence of functions. Some of them are performed internally, requiring only some table look ups to the global SRAM. Whereas other costly functions are carried out in a dedicated external accelerator, and therefore the required information has to be passed to the hardware block in the order that it performs its task. Once the processing of a packet has finished, the processor stores it in the packet memory (SDRAM).

Inside each multi-threading software processor a configurable number of contexts is available. A control mechanism cares for the context switches and implements itself a policy mechanism to decide which is the next thread to run. The instruction memory inside each processor is also divided in so many parts as contexts. The corresponding number of program counters point to the different context sections in which the instruction memory is divided.

The embedded software processors act as masters when sending a command request to the shared functional units inside the platform through the command bus. The target functional units store the requests in an input command fifo and process them in the sequence they arrived. The data bus is then shared by the functional units to perform the data transmissions indicated in the processed requests. Once a certain command has been completed, the target unit will signalise this to the initiator (the master who sent the command request first), sending it the corresponding event. In a multi-threading software processor, such events are the inputs which wake up the contexts, which were put to sleep after sending the related requests.

6.6 Case Study. Results and Evaluation

A subset of the TCP/IP packet functionalities inside an input packet processor is taken as example implementation (Figure 6.22). The packet is first picked up from the receive fifo within the interface unit. Further on, it is verified whether the length, the protocol version, and the checksum are the ones expected and, later, the modification of some fields takes place, as for example, the subtraction of the *time to live* field and the subsequent new checksum calculation. The packet is then classified, the selected policing algorithm is performed, and after obtaining the next hop IP address the packet is finally stored in the packet memory, SDRAM.

The single functions are modelled as SystemC modules with their respective input and output ports, and stored in the *computation nodes* library. The Functional model is achieved by collecting and linking them in the order defined by the specification. For the functional verification a test bench is built around, which is able to create and send packets with a parameterisable input-throughput and display the results, i.e. the packets arriving at the packet memory.

Towards the Architectural model for a first alternative, two instances of the Functional graph are created, and except for the classification function, the rest of the nodes are mapped to two different processing units, the software processors. The *classification* function of each Functional graph is mapped to another processing unit, the hardware block, in order to attempt the acceleration of this costly function. This node becomes thus a shared *computation node*. Finally, the *rfifo* and *sdram* functions are mapped to the shared interface unit and memory, respectively. These nodes become thus also shared *computation nodes*. The corresponding execution time values are extracted from the runtime library and annotated inside the nodes. In order to avoid access conflicts when

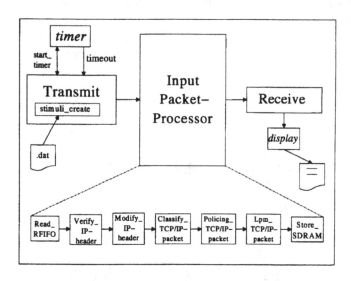

Figure 6.22. Case study: input packet processing

attempting to reach a shared *computation node* a policing mechanism, referred to as *shared element*, is added, which is accessed by the corresponding *access nodes* interleaved between the initiator and the shared *computation node*. This mechanism handles each request depending on the priority of the master which sent it.

The definition of the communication architecture fixes a command bus and a data bus as the communication media. Both the command bus and the data bus are bounded to the two software processors, to the hardware block and to the interface unit and memory unit. The bandwidth and frequency of both buses are provided as parameters to the communication media.

Each time a transfer between two functional units takes place two sequential external *communication nodes* have to be inserted, one mapped to the command bus and the other to the data bus. In the first case the software processors act as masters, sending a request to the selected shared functional unit through the command bus. Whereas in the second case the shared functional units act as masters, sending a write/read request to the initiator software processor through the data bus. These *communication nodes* solve the concurrent accesses to the buses and calculate the transfer duration. Moreover, they police the simultaneous accesses to the corresponding shared *computation node* or the access attempts while it is busy. For performing such tasks they access the corresponding Command Bus Arbiter, which take into account the priority of

188

the transfer initiator. These elements are introduced and customised according to the requirements.

The achieved Communication model for this partition alternative with two software processors and one hardware block accelerating the classification function is depicted in Figure 6.23.

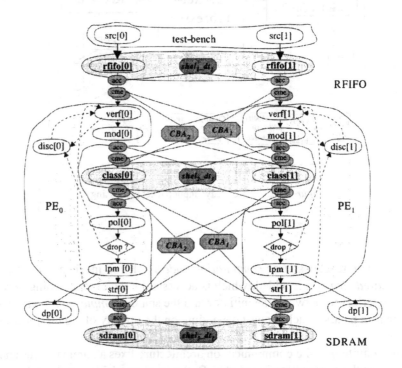

Figure 6.23. Mapping alternative: 2 PEs and 1 AC

Without requiring a re-building of the structure, a different mapping alternative for the input functionalities described above can be explored easily. In this case the *longest prefix match* algorithm for the calculation of the IP address of the next hop is mapped to a second hardware block, instead of implementing it inside the software processors. This single function is therefore sped up; however, a communication overhead is now added. Depending on the load of the buses, the duration of the transfers, and the load of the hardware block amongst others, this solution will lead to better or worse overall performance.

The simulation of the Communication model delivers the performance characteristics required for the evaluation of these partition alternatives. These results are compared with the throughput and simulation runtime values extracted from the cycle accurate model depicted in Figure 6.21. The comparison results are shown in the table 6.2, together with other architecture–partition alternatives consisting in two/six embedded processors and zero/one/two accelerators. The average throughput at the output side is measured stimulating the system with an input throughput of 915 Mbyte/s.

Table 6.2. Simulation runtime & throughput

		ASCSG (sec.)	Cycle Acc. (sec.)	Diff (%)	ASCSG (pack/sec.)	Cycle Acc. (pack/sec.)	Error rate (%)
2 PE							
	0 AC	2.18	6.66	67.27	288095	286108	-0.69
	1 AC	2.60	7.09	63.33	346651	341657	-1.46
	2 AC	3.07	8.07	61.96	443777	435351	-1.93
6 PE							
	0 AC	2.92	11.98	75.63	864765	858343	-0.74
	1 AC	3.34	12.58	73.45	1035220	1023630	-1.13
	2 AC	4.02	14.47	72.22	1325130	1304300	-1.59

Sun-W Ultra Sparc II, 360 MHz.

The analysis of the results demonstrates the considerable reduction of simulation runtime when applying the ASCSG method (around 70%), whereas the accuracy remains within an acceptable tolerance (around 1.5%). Even more important is the easy exploration of plenty partition alternatives, thanks to the low re-building effort and the automatised procedure.

6.7 Conclusions and Outlook

The increasing size, speed, and complexity of the latest designs require a higher level of abstraction than RTL. At the system level several trade offs in the design can be fixed which would be very costly at lower levels of abstraction. Therefore a system level estimation of the performance achieved by a certain architecture–partition alternative is very helpful for fixing such trade offs.

The performance estimation methodology presented in this chapter is thought to be integrated in a hardware–software co-design procedure, where an easy re-mapping of the functions onto the target architecture should be possible in order to enable the exploration of several partition alternatives. The current methodology achieves a fast simulation runtime while maintaining the accuracy within an acceptable tolerance. Furthermore, it makes the exploration of plenty of partition alternatives feasible by applying a graph whose structure has not to be re-constructed each time a new alternative is explored.

An Annotated SystemC Conditional Synchronisation Graph (ASCSG) is for this purpose built up, which comprises both the performance of each block and the characteristics of the communication architecture. The simulation of this graph shows its dynamic behaviour and delivers the pursued system performance estimation.

The implementation of the method is based on the system level language SystemC. Its support in terms of means of communication and rules for process activation, together with the fast simulation time achieved when modelling at transaction level, make this language very suitable for the implementation and automation of the method–procedure.

Chapter 7

SVE: A Methodology for the Design of Protocol Dominated Digital Systems

Robert Siegmund, Uwe Proß, Dietmar Müller

Department of Circuit and System Design
Chemnitz University of Technology, Germany

Abstract   In this article an efficient methodology is proposed for the SystemC based design of digital systems which are protocol dominated, such as communication networks based on serial protocols. In the development process of such systems a large amount of design effort has to be put into specification and implementation of protocol related hardware and software. The methodology presented here enables an abstract specification and simulation of complex frame-oriented protocols as well as synthesis of controller hardware from such specifications in the context of a SystemC system model. The protocol specification language is implemented as an extension library to SystemC 2.0. Using the USB 2.0 protocol as a modeling example we demonstrate the efficiency and benefits of the proposed methodology.

7.1   Introduction

Complex 'frame oriented' serial hardware communication protocols such as USB or FireWire, but also dedicated on-chip protocols, play an ever increasing role in the design of distributed digital systems for wireless and automotive applications, for highly integrated SoC and embedded systems. However, current design methodologies and tools have insufficiently addressed the high level specification, verification, design space exploration, and synthesis of such protocols in the system context. Hardware and software components related to protocol processing are in most cases heavily control dominated, and therefore their manual design and verification at a low level of abstraction is tedious and error prone.

A few synthesis oriented tools such as Synopsys Protocol Compiler [Synopsys, 1998] or PROGRAM [Öberg et al., 1996] exist which can transform an

W. Müller et al. (eds.),
SystemC: Methodologies and Applications, 191–216.

abstract high level protocol specification into RTL controller hardware. The drawback is that the specification means used are not part of a system description language, and therefore corresponding protocol specifications cannot be integrated into a HDL or SystemC based system model.

These arguments motivated the development of a SystemC library for high level specification and simulation of hardware communication protocols in the context of a SystemC model as well as synthesis from such specifications in order to tightly integrate such protocols into the SystemC based design of digital systems.

7.1.1   Related Work and Motivation

For the purpose of protocol specification formal specification languages such as ESTELLE and LOTOS [Turner, 1993] have been developed. ESTELLE models protocols in terms of a number of interacting extended finite state machines with each machine representing a communication endpoint. LOTOS is used to specify a protocol in terms of ordered event sequences. These languages have been mainly developed for the specification of OSI protocols and services at the system level and for automated formal checking of protocol properties.

SystemC 2.0 provides the high level modeling concept of *interfaces* and *channels* that originates from the SpecC language [Gajski et al., 2000], and enables a separation of computation and communication of a system specification. Channels provide high level communication links between behavioral specifications and encapsulate communication details such as the concrete signaling protocol or bus widths used. Behaviors access channels by means of interface method calls (IMC) such as read(), write() which are declared in channel interfaces. Communication protocols have to be modeled in terms of procedural descriptions that generate and consume signal sequences according to the protocol. These procedural descriptions implement the channel interface methods. When complex protocols such as USB are to be modeled using SystemC the problem persists that the procedural descriptions of protocol producer and consumer will contain extensively nested control structures which are tedious to code and hard to verify.

The goal of this work is therefore effectively to raise the abstraction level for protocol specifications in the context of a SystemC model. This goal is achieved through the implementation of a high level protocol specification language as an extension library to SystemC 2.0, named SVE. The name SVE originates from the *SuperVISE* design methodology [Hashmi and Bruce, 1995]. This methodology enabled a multi-level protocol specification based on an extension to VHDL, called VHDL+. Many of the concepts of this methodology have been realized in SVE.

SVE is a declarative language which lets designers specify hardware communication protocols in terms of patterns of valid signal sequences without the need to explicitly describe the protocol generating and consuming behaviors. Such protocol specifications are very concise and intuitive to read and can be simulated directly in the context of a SystemC model. Furthermore, hardware implementations for producer and consumer controller of the protocol can be synthesized from such a specification.

7.2   Specification of Data Communication Protocols

7.2.1   Communication Items

The basic element of a protocol specification using the SVE library is the *communication item* [Siegmund and Müller, 2001]. Communication items represent resources used for data transmission between behaviors at a specific level of abstraction. So, taking the USB protocol as an example, a communication

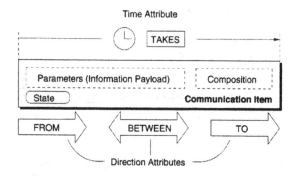

Figure 7.1.   Model of an SVE communication item

item may describe a complete USB frame, a data field within this frame or just a single bit of information in the serial USB bit stream. System modules (which can be either hardware or software components) exchange data and synchronize with each other through transmission of such communication items over abstract channels. Figure 7.1 visualizes the model of such a communication item.

An SVE communication item has a number of attributes which reflect typical properties of communication. The FROM, TO and BETWEEN attributes express that communication is directed. The direction is specified in terms of communication endpoints between which a particular item is transmitted. The TAKES attribute reflects the property that the transmission of an item consumes a certain amount of time. This time attribute can be set to an absolute value or to a time span, which introduces controlled non-determinism into protocol modeling. This is especially useful in early stages of design for constraining the design space without actually making a concrete implementation decision. Finally, items can have parameters describing their information content or payload transmitted with this item.

```
SV_MESSAGE(Packet) {

    // item parameters
    SV_Param<sc_uint<8> > addr;
    SV_ParamArray<sc_uint<8>,8> data;

    // item constructor
    SV_MESSAGE_CTOR(Packet) {

        SV_FROM << "PacketSender";
        SV_TO   << "PacketReceiver";
        SV_TAKES(100,2340);

        SV_PARAMETER(addr);
        SV_PARAMETER(data);

        // optional item composition
        SV_COMPOSITION(...);
    }
};
```

Figure 7.2. Specification of a communication item using the SVE library

In the SVE library two different types of communication items are defined, describing different levels of communication abstraction. The most abstract item is the frame item. Frame items describe a packet oriented data transfer between system modules, and are subdivided into transaction and message items. Transaction items are used for the specification of a multi-directional information transfer, e.g. protocols that involve a turnaround in the direction of transfer such as handshaking protocols. Message items specify a unidirectional information transfer between system modules. An example specification of a frame item using the SVE library is given in Fig. 7.2. This item specifies a packet containing a byte address and eight bytes of data.

An item specification consists of a declaration part which defines the item parameters and an item constructor. Parameters can be scalars or arrays, and are defined as SV_Param<type> or SV_ParamArray<type>, respectively. The template type argument can be any C++, SystemC, or user defined type. For a parameter array, the second template argument defines the size of that array. In Fig. 7.2 the frame item Packet has two parameters, a scalar address value and an array of eight data values. In the item constructor the direction and timing attributes are set, and all variables of type SV_Param that represent item parameters are registered as such. The order in which they are registered is used for positional mapping of actual to formal parameters when an item is instantiated. Other variables of type SV_Param that are declared in an item but not registered as a parameter serve as local variables used to store the item state. Finally, for transaction and message items a transmission protocol can be declared in form of a composition of other items (see section 2.2 of this article).

```
SV_PHYMAP(MapBitToWire) {

    SV_Param<bool> XD_val;
    SVSignalRef<bool> XD;

    SV_PHYMAP_CTOR(MapBitToWire) {

    SV_PARAMETER(XD_val);
    SV_ASSOCIATE(XD - XD_val);
    }
};
```

Figure 7.3. Specification of a PHYMAP item using the SVE library

The second type of item provided with the SVE library is the PHYMAP item. This item provides a means for mapping of abstract transactions and messages and their parameter values to sequences of physical signal states. Fig. 7.3 shows an example for a PHYMAP item specification. In this example the parameter XD_val is mapped to the physical signal XD. In the declarative part of a PHYMAP item, parameters as well as references to the SystemC signals to which the parameter values are to be mapped are declared. Finally, in the PHYMAP item constructor the referenced signals are associated with the item parameters.

7.2.2 Item Compositions

Frame oriented protocols are described in SVE through composition of items to new items. An item composition specifies a schedule in which a set of less abstract items that constitute a more abstract item, are transmitted.

Before going into detail it has to be pointed out that SVE protocol specifications given as item compositions are *purely declarative* specifications and describe a protocol in terms of legal item sequences for a protocol. Owing to the declarative nature of SVE protocol specifications they cannot directly be run on a computer. Instead, in the elaboration phase of a simulation run the SVE kernel will derive two distinct, executable behaviors from such a protocol specification:

196

▷ The *decompositional behavior* defines how a communication item is decomposed over time into a set of constituting lower level items according to its protocol. It defines, furthermore, how the set of item parameters are mapped to the formal parameters of these lower level items. From the implementation viewpoint it specifies the behavior of the *protocol producer* controller;

▷ The *compositional behavior* corresponds to the *inverse* decompositional behavior and defines how a set of constituting lower level items are assembled over time according to the protocol in order to generate a higher level item. It defines, furthermore, how the information payload contained in the parameters of the assembled item is reconstructed from the parameter values of the constituting items. From the implementation viewpoint it specifies the behavior of the *protocol consumer* controller.

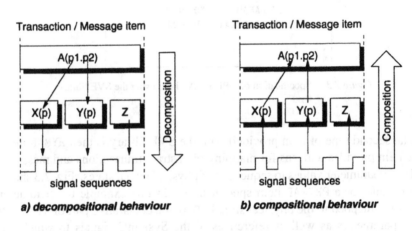

Figure 7.4. Reversibly executable protocol specification

Because compositional and decompositional behaviors are the inverse of each other (see Fig. 7.4) an SVE protocol specification is said to be *reversibly executable.*

In order to obtain a declarative, reversibly executable protocol specification, communication items have to be composed by means of predefined composition schemes. A total of four different schemes are provided with the SVE library, which can be arbitrarily nested. These composition schemes are depicted in Fig. 7.5. The SERIAL and PARALLEL schemes schedule a number of items for sequential or parallel transmission, respectively. In a serial composition, items are transmitted back to back while in a parallel composition the transmission of all items starts at the same time. Figure 7.6 shows the syntax of these composition schemes. The () operator is required for each item instantiated in a

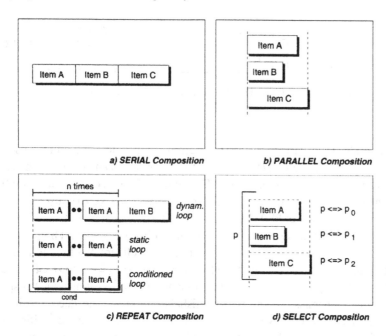

Figure 7.5.   Item composition schemes provided with the SVE library

composition scheme in order to build the appropriate simulation data structures during execution of the item constructor. The decompositional and compositional behaviors of these schemes are straightforward. In the SERIAL scheme the items are generated and matched in sequence, and in the PARALLEL scheme they are generated and matched concurrently. If an item cannot be matched at the time it is expected the compositional behavior generates a protocol error which is displayed in a simulation.

```
SV_SERIAL(
    A(),B(),C()
);
```

```
SV_PARALLEL(
    A(),B(),C()
);
```

Figure 7.6.   Syntax of the SERIAL and PARALLEL composition schemes

The two more complex composition schemes provided with the SVE library are the REPEAT and SELECT schemes. The first variant of the REPEAT scheme specifies a statically bounded loop for repeated transmission of an item. In this scheme a loop variable of type SV_Param, which is named i in the example, and the lower and upper bounds of the loop must be specified. The decompositional and compositional behaviors of this scheme correspond to those of the serial

```
SV_REPEAT(
    i,7,0,A(data[i])
);
```

```
SV_REPEAT(
    i,A()
);
```

```
SV_REPEAT(
    i,N,A(data[i]),B()
);
```

Figure 7.7. Syntax of the variants of the REPEAT composition scheme

scheme. The loop variable holds the current loop count and can be used to select elements from a parameter array that is passed as an actual parameter to the item instance in the loop body. The static REPEAT scheme can therefore efficiently describe a reversible parallel to serial and vice versa data conversion.

The second variant of REPEAT scheme specifies a repeated transmission of an item, which is terminated as soon as a condition becomes false. This condition must be evaluated separately with the aid of a user defined C++ function and the result assigned to a variable of type SV_Param<bool>. This form of REPEAT scheme is mostly used to control the transmission of an item (e.g., suspend and resume transmission) through handshake signals. The decompositional behavior of this scheme would generate item A as long as the condition i is true. The compositional behavior would try to match item A as long as the condition is true. In case the condition is true and item A could not be matched a protocol error would be generated during simulation.

Finally, there is a third variant of the REPEAT scheme. This scheme declares a repeated transmission of an item that is terminated by the transmission of an item distinct from the first one. The decompositional behavior of this scheme would generate item A a number of N times, followed by generation of item B. The compositional behavior, however, is this that every time item A is matched the loop variable is incremented starting from zero until item B is matched, so that after a match of item B the loop count parameter contains the number of matched items A. If neither item A nor B can be matched a protocol error is generated. An application for this scheme is the modeling of serial protocols that contain a variable number of data fields delimited by an end field, but there is no explicit information on the number of data fields in the protocol stream. Here the compositional behavior automatically extracts both the transmitted data and data length information from the stream.

The SELECT scheme provided by the SVE library is used to selectively transmit a particular item chosen from a set of alternative items. The selection is based on the value of an item parameter. In Fig. 7.8 the parameter p is used to choose one from three alternative items. The SELECT scheme has a template argument that is used to specify the type of the selection parameter. The decompositional behavior of the SELECT scheme generates the item specified in the branch that is selected by the parameter p. The compositional behavior of this scheme (left example in Fig. 7.8) would try to match one of the specified

```
SV_SELECT<sc_uint<8> >(           SV_SELECT<sc_uint<8> >(
    p,                                SV_MATCH(p),
    sc_uint<8>(0) <= A(),             sc_uint<8>(0) <= A(),
    sc_uint<8>(1) <= B(),             sc_uint<8>(1) <= B(),
    sc_uint<8>(2) <= C()              sc_uint<8>(2) <= C()
);                                );
```

Figure 7.8. Syntax of the variants of the SELECT composition scheme

alternative items, and in the case of a successful match it would set parameter p
to the corresponding value and otherwise generate a protocol error. The compositional behavior can be modified such that for parameter p a certain value is
expected, and hence only the item selected by this value is to be matched (right
example in Fig. 7.8). The fact that the selection parameter needs to match a
certain value instead of being established is expressed with the SV_MATCH prefix.

7.2.3   Embedding Sequential Behaviors in Protocol Specifications

As described in section 2.1, communication items may have local variables
which are used to store an item state. These variables are mostly used as selection or loop bound parameters in a SELECT or REPEAT scheme, when a protocol
with data dependencies is to be modeled. However, the SVE language elements
presented so far provide no means to set or modify the contents of local variables. For this purpose the SVE library enables the embedding of sequential
behaviors in protocol specifications. A sequential behavior is a C++ method defined in the item specification, and embedded in the item composition using the
SV_SEQBEHAVIOUR statement (see Fig. 7.9). In order to ensure the reversibility
of a protocol specification there are some restrictions to the implementation of
sequential behaviors. In particular must a sequential behavior execute in zero
time (e.g. must not contain wait statements). It furthermore must not modify
item parameters. The item specification in Fig. 7.9 uses an embedded sequential
behavior compute_crc to compute the checksum over a data field and to store
the checksum value into variable crc, which is then passed as actual parameter
to item CRC.

7.2.4   Interface Units

Communication items are allocated in a design unit that is called the *interface
unit*. An interface unit contains a collection of references to all items and
physical signals related to a protocol specification. An example specification
of an interface unit for a protocol named XProtocol, which makes use of the

```
SV_MESSAGE(Packet) {

  SV_ParamArray<sc_uint<8>,8> data; // parameter
  SV_Param<sc_uint<16> > crc;       // local variable

  void compute_crc() {...}          // embedded behavior
                                    // compute crc from data
  ...
  SV_COMPOSITION(
    SV_SERIAL(
      Data(data),
      SV_SEQBEHAVIOUR(compute_crc),
      CRC(SV_MATCH(crc)))
  );
};
```

Figure 7.9. Embedding of sequential behaviors in item compositions

previously defined items `Packet` and `MapBitToWire`, is given in Fig. 7.10.
The implementation of the interface unit constructor needs to be placed into

```
// file XProtocol.h
SV_INTERFACE(XProtocol) {

    SVMessageRef Packet_M;
    SVPhymapItemRef MapBitToWire_P;
    SVSignalRef<bool> XD;

    SV_INTERFACE_CTOR_DECL(XProtocol);
};

SV_MESSAGE(Packet) {...}
SV_PHYMAP(MapBitToWire) {...}

// file XProtocol.cc
SV_INTERFACE_CTOR(XProtocol) {

    SV_BETWEEN << "PacketSender" << "PacketReceiver";

    SV_ALLOCATE_ITEM(Packet, Packet_M);
    SV_ALLOCATE_ITEM(MapBitToWire, MapBitToWire_P);
    SV_REGISTER_SIGNAL(XD);
};
```

Figure 7.10. Specification of an interface unit

a separate file, as shown in Fig. 7.10. The constructor contains the code for allocation of all items and registration of external signals. Allocated items are then accessible in compositions of other items via their references in the interface unit. In the constructor also the set of communication endpoints between items can be transferred are defined.

7.2.5 Specification of Communicating Behaviors

The previous sections gave an introduction on how protocols are specified using the SVE library in terms of communication items and item compositions. In order to integrate such a protocol specification into a SystemC system model and let behaviors communicate by means of item transfers, an instance of an interface unit needs to be created. An instance of an interface unit is called a *channel* and will be connected to the behaviors via ports. Behaviors then use

```
SC_MODULE(Behavior_A) {
    SV_InterfacePort<XProtocol> T;

    void transmit_packet() {
      T.sv_send(T->Packet_M(addr,data)); }

    SC_CTOR(Behavior_A) {
      T.SV_IMPLEMENTS("PacketSender"); }
};

SC_MODULE(Behavior_B) {
    SV_InterfacePort<XProtocol> T;

    void receive_packet() {
      T.sv_receive(T->Packet_M(addr,data)); }

    SC_CTOR(Behavior_B) {
      T.SV_IMPLEMENTS("PacketReceiver"); }
};

// instantiate behaviors and interface unit
Behavior_A A("A");
Behavior_B B("B");
SV_Channel<XProtocol> C("C");

// connect behaviors via a channel
A.T(C); B.T(C);
```

Figure 7.11. Specification of two behaviors communicating over a channel

the predefined SVE methods `sv_send()` and `sv_receive()` in order to send and receive items via these ports. For each port it must be specified in the module constructor, which communication endpoint from the ones given in the interface unit it implements. A port must implement exactly one communication endpoint. The code example in Fig. 7.11 shows the specification of two SystemC modules which are connected via an SVE channel and which communicate using the Packet message item.

7.2.6   Automatic Translation Between Levels of Communication Abstraction

In the case in which for a SVE communication item a composition is declared the sv_send() method will invoke the decompositional behavior whilst the sv_receive() method will invoke the compositional behavior of this item. If an item composition is considered as an implementation of this item at a lower level of abstraction, the execution of the decompositional behavior will generate all lower level representations of this item in terms of sequences of lower level items and, finally, sequences of signal states. Accordingly the compositional behavior of this item will try to assemble all higher level representations from the generated sequence of lower level items. This fact can effectively be used to perform an automatic translation between levels of communication abstraction. It is especially useful in the case in which two modules of different abstraction need to be connected together for a system simulation. This situation is depicted in Fig. 7.12. Here a module given as an abstract behavioral model (ABM)

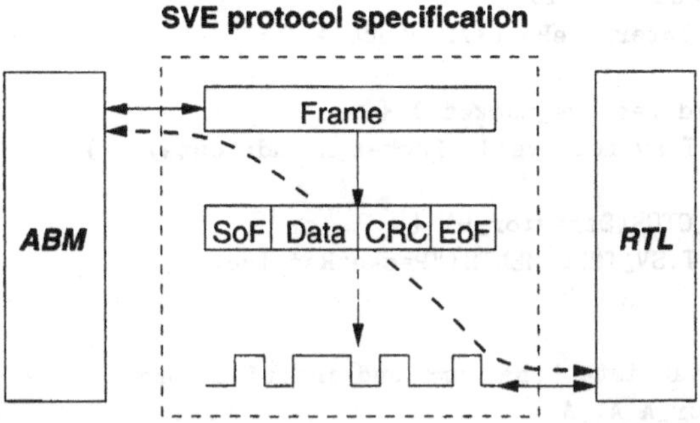

SVE protocol specification

Figure 7.12.   Automatic translation between levels of communication abstraction using an SVE protocol specification

needs to be connected to another module implemented as an RTL model. Both modules talk to each other using a frame oriented protocol such as USB, but the RTL model has a detailed pin level interface whilst the abstract model is capable only of sending and receiving abstract frames. This situation may, for example, occur when an IP core is to be simulated in the context of an abstract system model. By executing the decompositional and compositional behavior of the SVE protocol model frames are automatically translated top down and bottom up from their abstract representation into a signal sequence and vice versa so that both modules are able to communicate.

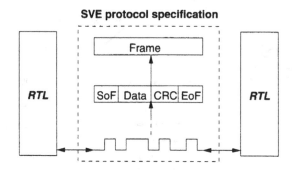

Figure 7.13. Using a SVE protocol specification for protocol checking

Another application of an SVE protocol model is depicted in Fig. 7.13. Here, both modules are implemented as RTL models, so the decompositional behavior of the protocol specification is not exploited. However, through exercising the compositional behavior the RTL signal sequences transmitted between the modules can be checked if they conform to the specified protocol. Therefore SVE protocol specifications can also serve efficiently as protocol checkers if they are attached to a communication medium.

7.3   An SVE Model of the USB 2.0 Protocol

7.3.1   The USB 2.0 Protocol

The Universal Serial Bus protocol version 2.0 [Compaq et al., 2000] has been chosen owing to its wide application and sufficient complexity in order to demonstrate the modeling capabilities of the SVE library. The USB protocol is a packet oriented protocol, which uses packets of variable contents and variable length. As shown in Fig. 7.14, the components of a USB network are connected in a tiered star topology. The center of each star is a hub. The root hub is called USB host and comprises the USB host controller. This controller is the interface to the host computer system, which is usually a PC. Hubs are used to provide additional attachment points to the USB and allow connecting functions running in different speed modes. A function is a USB device that provides a certain service, such as a printer or scanner, to the USB system. For USB 2.0 three different speed modes are defined in order to achieve the optimal trade off between performance requirements and implementation costs. USB uses a two-wire bit serial transmission where each bit of information is encoded in a

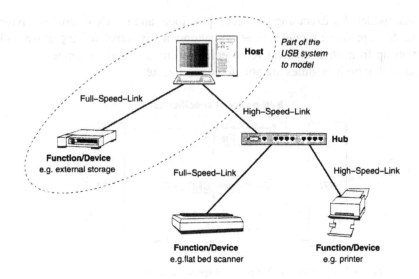

Figure 7.14. Topology of an USB system

Table 7.1. Speed modes used in USB 2.0

	Low Speed	Full Speed	High Speed
Signal rate	1,5 Mbit/s	12 Mbit/s	480 Mbit/s
Data rate	10 – 100 kbit/s	500 kbit – 10 Mbit/s	25 – 400 Mbit/s
Applications	Interactive devices, Mouse, Keyboard	Communication, Printer, Microphone	External storage, Video, Imaging

pair of voltage levels on the two wires. Each one of the speed modes shown in Table 7.1 has a distinct mapping of bit information to a pair of voltage levels on the USB wires (see also section 3.2).

7.3.2 The USB Protocol Model

In this section the modeling of the USB protocol based on SVE items and item compositions is described. Owing to the complexity of the USB protocol some aspects of the protocol had to be chosen that include USB packet selection, bit stuffing, NRZI (Non Return to Zero Inverted) encoding and bus state mapping. For these aspects of the protocol an efficient modeling solution using the SVE library will be presented.

Packet Selection. USB transfers data between host and function in units of packets. USB packets have a variable format. The format depends on the packet type, which is specified by a packet identifier field (PID) at the beginning of the packet. The USB protocol defines a total of 16 different packet types, which are divided into token packets, data packets, handshake packets, and special packets. A simplified overview of the USB packet structure is shown in Fig. 7.15. The SVE specification of a USB packet is shown in Fig. 7.16. A USB

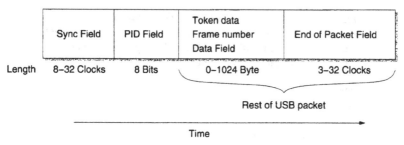

Figure 7.15. Structure of an USB packet

packet is described as an SVE item of type message. The message specification uses the serial composition scheme for the description of the packet structure according to Fig. 7.15. The packet type is selected by means of the parameter pid in conjunction with a SELECT composition. The decompositional behavior of this message would generate items SyncField and PIDField, followed by one of the token, data, packet or handshake items, depending on parameter pid. In the compositional behavior of this message the value of parameter pid will be established at the time message PIDField has been matched. Since now the type of item that must follow next is predetermined by the value of this parameter, the SV_MATCH keyword is applied to the parameter in the SELECT composition to account for the fact that the item related to the pid is expected next in the protocol stream.

Bit Field of Fixed Length. The USB PID field is a bit field of fixed length. Fig. 7.17 illustrates the format of the PID field. The eight-bit PID value is transmitted from LSB to MSB. The corresponding SVE description is shown in Fig. 7.18. The composition of this message is quite simple. The static repeat composition is used to describe the PID field in form of a serial bit stream which is formatted LSB to MSB. During execution of the decompositional behavior, the value of the PID field is transmitted bit by bit with item LogicalBit. The compositional behavior would try to match item LogicalBit eight times. It would then assign the bit value received with this item to the corresponding bit

```
SV_MESSAGE(USBPacket) {

  SV_Param    <sc_uint<8> > pid;    // Packet type
  SV_Param    <sc_uint<10> > datalen; // data field length
  SV_Param    <sc_uint<16> > addr;  // endpoint address
  SV_ParamArray<sc_uint<8>,8> data; // data field
  ...
  SV_COMPOSITION(
    SV_SERIAL(
      SyncField_M(),
      PIDField_M(pid),
      SV_SELECT <sc_uint<8> > (
        SV_MATCH(pid),
        // Token packets
        sc_uint<8>(PID_IN)  <= TokenPacket_M(pid,addr),
        ...
        // Data packets
        sc_uint<8>(PID_DATA0) <= DataPacket_M(pid,data,
                                              datalen),
        ...
        // Handshake packets
        sc_uint<8>(PID_ACK) <= HandShakePacket_M(pid),
        ...
        // Special packets
        sc_uint<8>(PID_PRE) <= SpecialPacket_M(pid,addr),
        ...
  )
  );
};
```

Figure 7.16. Specification of a USB 2.0 packet

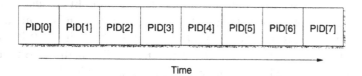

PID[0]	PID[1]	PID[2]	PID[3]	PID[4]	PID[5]	PID[6]	PID[7]

Time →

Figure 7.17. Format of an USB PID field

```
SV_MESSAGE(PIDField){
    SV_Param<sc_uint<8> > pid;
    SV_Param<sc_uint<8> > idx;

    SV_COMPOSITION(
        SV_REPEAT(
            inxx,0,7,LogicalBit_M(pid[idx]))
    );
};
```

Figure 7.18. Specification of the USB PID field

in the PID field. After eight successful matches of the LogicalBit item the value of the PID is reconstructed.

Bit Stuffing. USB does not use an explicit clock line to transmit synchronization information between host and function. Instead, the communication endpoints use autonomous phase locked loop based clock generators, which need to be synchronized in certain intervals. The synchronization information can be obtained from transmitted data as long as a sufficient number of signal transitions take place on the USB wires in a certain time interval. In order to ensure sufficient signal transitions, bit stuffing is applied to the USB data stream. A zero bit is inserted by the sender respectively removed by the re-

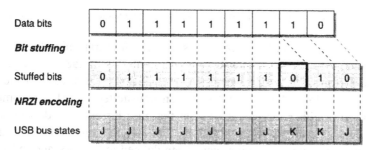

Data bits	0	1	1	1	1	1	1	1	0
Bit stuffing									
Stuffed bits	0	1	1	1	1	1	1	0	1	0
NRZI encoding										
USB bus states	J	J	J	J	J	J	J	K	K	J

Figure 7.19. USB bit stuffing and NRZI encoding

ceiver after six consecutive ones bits in the data stream. In conjunction with
NRZI encoding (see the following section) a signal transition is guaranteed on
the USB wires at least on every seventh transmitted bit. Fig. 7.19 illustrates
the usage of bit stuffing in conjunction with NRZI coding in the USB protocol.
The composition of the bit stuffing message is shown in Fig. 7.20.

```
SV_MESSAGE(LogicalBit) {

  SV_Param<bool> bitval;
  SV_Param<sc_uint<3> > bit_count; // ones counter

  void clearBitCnt() { bit_count=0; }
  void incrBitCnt() { bit_count++; }
  ...
  SV_CONSTANT(zero_bit, 0);

  SV_COMPOSITION(
    SV_SERIAL(
      BusStateNRZI_M(bitval),

      SV_SELECT <bool> (
        SV_MATCH(bitval),
        bool(0) <= SV_SEQBEHAVIOUR(clearBitCnt),
        bool(1) <= SV_SEQBEHAVIOUR(incrBitCnt)),

      SV_SELECT <sc_uint<3> > (
        SV_MATCH(bitCount),
        sc_uint<3>(6) <= SV_SERIAL(
                           BusStateNRZI_M(zero_bit),
                           SV_SEQBEHAVIOUR(clearBitCnt)),
        SV_OTHERS     <= SV_NULL)
    )
  );
};
```

Figure 7.20.   Specification of the USB logical bit message which implements bit stuffing

The decompositional behavior of this message first transmits the actual bit
value contained in the parameter bitval using message BusStateNRZI. This
message performs the NRZI encoding of the USB bit stream and is described
in the next section. After the bit has been transmitted, a SELECT composition is used to either increment or reset the ones bit counter, depending on
the value of the currently transmitted bit. For incrementing and resetting this
counter two sequential behaviors have been defined in the message in form of
item methods and have been embedded into the message composition using
the SV_SEQBEHAVIOUR keyword. A second SELECT composition tests the bit

counter if a total of six ones bits have been transmitted in sequence. In this case a stuffing zero bit will be transmitted next using message BusStateNRZI and the bit counter will be cleared.

The compositional behavior first attempts to match the message BusState-NRZI and in the case of success it establishes the bit value transmitted with this message in parameter bitval. Depending on this value, the ones' bit counter is then incremented or cleared as in the decompositional behavior. However, the difference is that if six consecutive ones bits have been received, the compositional behavior next attempts to *match* the message BusState-NRZI. This message must transmit the stuffed zero bit. If this message can be matched successfully, then the bit stream transmitted so far conformed to the bit stuffing rule. Otherwise the compositional behavior would indicate a protocol error. The specification of the bit stuffing message shows how efficient quite complex compositional and decompositional behaviors can be expressed using the declarative SVE syntax.

NRZI Encoding. As mentioned before, USB applies NRZI encoding to the stuffed bit stream. An NRZI encoded bit stream consists of the symbols J and K, representing two of the four possible USB bus states. As depicted in Fig. 7.19, a ones' bit is represented by no bus state change, whilst a zero bit is represented by a transition of the bus state from J to K or from K to J.

Fig. 7.21 shows the specification of the USB message which implements NRZI encoding. Nested SELECT compositions are used to translate the value of the message parameter databit into the appropriate consecutive pair of JK symbols. These symbols will be generated by the corresponding messages BusStateJ and BusStateK, respectively. In both the decompositional and compositional behaviors, with the outer SELECT composition the appropriate encoding of the next bit value is chosen depending on the last bus state, which is stored in the parameter lastBusState. The decompositional behavior of the inner SELECT compositions will generate the next bus state J or K, depending of the value of parameter databit, and set the parameter lastBusState accordingly. The compositional behavior of the inner SELECT compositions, however, will try to match either a J or K symbol and set the selection parameter to the corresponding bit value. Therefore the compositional behavior effectively reverses the NRZI encoding through translation of consecutive pairs of JK bus states back into the original bit values. This code example again demonstrates how quite complex serial data stream formats and protocols can be efficiently specified using the declarative syntax of SVE.

Bus State Mapping. In a USB cable there are, amongst others, two wires, called D+ and D-, which represent the physical medium used for data transmission. Each one of the USB speed modes uses its own mapping of the USB

```
SV_MESSAGE(BusStateNRZI) {

    enum BusStateType {J,K,SEO,IDLE}; // USB bus states
    SV_Param<bool> databit;           // bit value to be encoded
    SV_Param<BusStateType> lastBusState; // last bus state

    void setlastBusStateJ() {lastBusState=J;}
    void setlastBusStateK() {lastBusState=K;}
    ...
    SV_COMPOSITION(

        SV_SELECT<BusStateType> (
          SV_MATCH(lastBusState),

          BusStateType(J) <=
            SV_SELECT<bool>(databit,

                bool(0) <= SV_SERIAL(
                           BusStateK_M(),
                           SV_SEQBEHAVIOUR(setlastBusStateK)),

                bool(1) <= SV_SERIAL(
                           BusStateJ_M(),
                           SV_SEQBEHAVIOUR(setlastBusStateJ))
            ),
          BusStateType(K) <= ...
    );
};
```

Figure 7.21. Specification of the USB message which performs NRZI encoding

bus states to voltage levels on the USB wires D+ and D-. The bus state mapping defined in the USB 2.0 specification [Compaq et al., 2000] is shown in Table 7.2. Fig. 7.22 shows the specification of message item BusStateJ. Item BusStateK is specified in an analog fashion.

The composition of this message is quite simple and contains just a SELECT composition of three instances of the PHYMAP item USBPhy with different actual parameters. This PHYMAP item forms the interface to the physical USB wires. The actual parameters specify the appropriate signal voltage levels for the desired bus speed mode. The selection of these voltage levels is based on a variable BusSpeed, which is in this case not an item parameter but a global variable defined in the interface unit. The last thing that remains to present for a complete protocol model is the specification of the USB Phymap item. This specification is shown in Fig. 7.23. The item specification contains the references to the two USB wires D+ and D-, which are associated with the item

Table 7.2. Mapping of the bus states to voltage values on the USB wires

Bus state	J		K		SE0	
USB wires	D+	D-	D+	D-	D+	D-
High-Speed	V_{HSOH}	V_{HSOL}	V_{HSOL}	V_{HSOH}	-	-
Full-Speed	V_{OH}	V_{OL}	V_{OL}	V_{OH}	V_{OL}	V_{OL}
Low-Speed	V_{OL}	V_{OH}	V_{OH}	V_{OL}	V_{OL}	V_{OL}

$V_{HSOH} = 360 \ldots 440 \, \text{mV}$ High Speed High level
$V_{HSOL} = -10 \ldots 10 \, \text{mV}$ High Speed Low level
$V_{OH} = 0 \ldots 0.3 \, \text{V}$ Full/Low Speed High level
$V_{OL} = 2.8 \ldots 2.6 \, \text{V}$ Full/Low Speed Low level

```
SV_MESSAGE(BusStateJ) {

    // Voltages on the USB wires in millivolts
    SV_CONSTANT(HSOL, 10);
    SV_CONSTANT(HSOH, 440);
    SV_CONSTANT(OH, 2800);
    SV_CONSTANT(OL, 300);
    ...
    SV_COMPOSITION(
      SV_SELECT <BusSpeedType> (
        SV_MATCH(BusSpeed),
          BusSpeedType (HIGH_SPEED) <= USBPhy_P(HSOH,HSOL),
          BusSpeedType (FULL_SPEED) <= USBPhy_P(OH,OL),
          BusSpeedType (LOW_SPEED) <= USBPhy_P(OL,OH))
    ));
};
```

Figure 7.22. Specification of the USB message `BusStateJ`

parameters using the SV_ASSOCIATE keyword. Depending on whether the decompositional or the compositional behavior of this item is executed, the values of parameters DPval, DNval are assigned to the signals DP, DN or vice versa.

7.3.3 USB System Model

The protocol description developed in the previous section has been used to model a USB communication link comprising a host and a device, as depicted in Fig. 7.14 [Pross, 2002]. USB host and device are connected via an SVE channel encapsulating the USB protocol and communicate by sending and receiving the previously described USBPacket item. This system model has been simulated at various protocol abstraction levels. (see Fig. 7.24).

```
SV_PHYMAP(USBPhy) {

  SV_Param<int> DPval; // D+
  SV_Param<int> DNval; // D-

  // USB wires
  SV_SignalRef<int> DP; // D+
  SV_SignalRef<int> DN; // D-

  SV_PHYMAP_CTOR(USBPhy) {

    // associate the parameters with the USB signals
    SV_ASSOCIATE(
           DP - DPval,
           DN - DNval);
  }
};
```

Figure 7.23. Specification of the USB PhyMap item

The abstraction level of the protocol has been varied through removal of lower level items from the protocol specification. Fig. 7.26 shows the relative simulation time of the system model for various protocol abstraction levels.

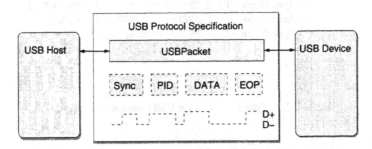

Figure 7.24. Abstract model of a USB communication link

In the simulation, the transmission time of a USB token packet was measured for a USB protocol specification that describes packet level only, a specification that describes packet and field level and a specification that describes the protocol from packet to bus state level. The experiment showed that the simulation time increases rapidly by refining the protocol specification. A refinement down to field level increases the simulation time by a factor of 3. When the USBPacket item is specified down to the detailed bus state level, simulation times are about 12 times higher compared to a simulation using the abstract

`USBPacket` without composition. This fact can be used to speed up system simulations by removing unnecessary communication details if, e.g., only the functionality of host and device is to be examined.

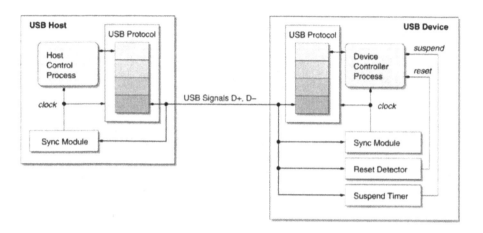

Figure 7.25. Implementation of the USB communication link to simulate clock recovery

In the next step host and device modules were refined in order to simulate clock recovery from the USB data stream (see Fig. 7.25). Clock recovery is needed to synchronize the sampling clock generators in host and device. For this purpose corresponding synchronization modules that generate the sampling clock for the USB wires have been added to host and device modules. Two instances of the USB protocol specification are used to serve as protocol generator and protocol consumer, respectively. Host and device are connected via the USB wires D+, D– to which also the synchronization modules of host and device are connected (Fig. 7.25). The relative simulation time for this model is also shown in Fig. 7.26 and is about 86 times higher compared to the abstract model of the USB communication link.

7.4 Synthesis from SVE Protocol Specifications

SVE protocol specifications are fully synthesizable into hardware implementations of corresponding protocol controllers. This is, in fact, another advantage over SystemC 2.0 interfaces and channels for which currently no automated synthesis into hardware exists. Synthesizing an SVE protocol specification means that hardware implementations for *both* the decompositional and the composi-

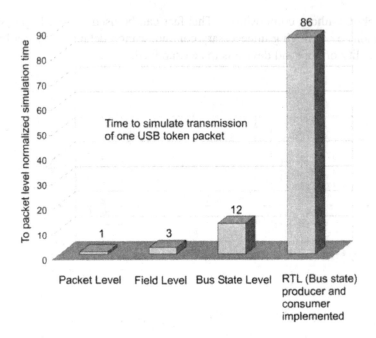

Figure 7.26. Simulation time of the USB protocol at different abstraction levels

tional behavior are generated in form of a pair of protocol controllers. These two protocol controllers act as protocol generator and protocol consumer, respectively. The principle of controller synthesis is depicted in Fig. 7.27.

A tool named COSYNE (*C*ontroller *Syn*thesis *E*nvironment) has been implemented with about 15000 lines of C++ code, which analyzes an SVE protocol description, performs the controller synthesis and generates implementations of the protocol controllers in the form of synthesizable SystemC RTL models [Siegmund and Müller, 2002]. The input to this tool is a complete SystemC system model comprising a number of module specifications that are connected by means of an SVE channel. This channel encapsulates the communication protocol to be used. The SystemC RTL code of the two protocol controllers resulting from the synthesis process are integrated by the COSYNE tool into the module specifications. Finally, the signal level interfaces of the interacting controllers are connected by wires. Provided that the module specifications are synthesizable, the synthesis output of the COSYNE tool is a synthesizable SystemC model of the system with communication refined to the signal level.

COSYNE was used to synthesize SystemC 2.0 RTL code for producer and consumer controllers for the USB 2.0 protocol. Producer and consumer controllers comprise the functionality of the USB protocol packet assembling/disassembling engine including packet selection, bit stuffing, and NRZI encoding

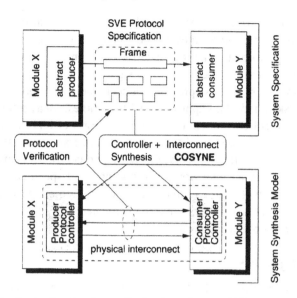

Figure 7.27. Synthesis of protocol controller hardware from an SVE protocol specification

and decoding, respectively. The generated SystemC RTL models were synthesized into a gate level implementation for a $0.13\mu m$ technology and for a clock frequency of 62.5 MHz, using Synopsys Design Compiler. Table 7.3 shows the synthesis results for the controller implementations.

Table 7.3. Synthesis results for the USB 2.0 protocol specification

	Producer Controller	*Consumer Controller*
FSM State Count	298	438
Clock Frequency (MHZ)	62.5	62.5
Register Count	145	223
Total Area (Gates)	1368	1718

7.5 Summary

An efficient methodology for integration of abstract specifications of hardware communication protocols into SystemC 2.0 system level models has been presented. The methodology is based on a declarative protocol specification language (SVE) which has been implemented as an extension library to SystemC 2.0. The advantages of declarative SVE protocol specifications have been demonstrated by means of a model of the USB 2.0 protocol. Such specifications are concise and intuitive to create and understand by designers, and effectively

216

raise the level of abstraction for protocol modeling. Because such specifications are reversibly executable they can be used as protocol generators, protocol checkers or can perform automatic translation between levels of communication abstraction in a mixed abstraction level design.

Furthermore, the synthesis approach for such declarative protocol specifications has been sketched. This approach enables synthesis of a protocol generator as well as a protocol consumer controller in the form of RTL models from the *same* protocol description and their automated integration into the communicating behavioral specifications. Synthesizing two controller implementations with completely distinct behaviors from one protocol specification is a novel approach in high level synthesis that does not only tremendously raise the design productivity but also ensures that the generated implementations conform to the initial protocol specification.

Acknowledgments

We would like to thank the German National Merit Foundation for partly sponsoring this work.

Chapter 8

Object Oriented Hardware Design and Synthesis Based on SystemC 2.0

Eike Grimpe,[1] Wolfgang Nebel,[2] Frank Oppenheimer[1], Thorsten Schubert[1]

[1] *OFFIS Research Institute, Oldenburg, Germany*

[2] *Carl v. Ossietzky University of Oldenburg, Oldenburg, Germany*

Abstract   In this article we will briefly discuss the good prospects that SystemC opens for high level, and especially object oriented, hardware modeling. But we will also highlight the fundamental problems that SystemC's higher level language constructs represent for hardware synthesis. We will further present how, based on SystemC, object oriented hardware modeling and a clear synthesis semantics can be combined, leading to an extended SystemC/C++ synthesis subset. And last, we will outline two synthesis strategies that allow direct and automated hardware synthesis from object oriented hardware specifications based on the extended synthesis subset mentioned above.

8.1   Introduction

Object orientation [Booch, 1993] currently is the dominant paradigm in software engineering. The basic idea of building a system by decomposing it into a set of associated objects can be found in abstract methodologies, system analysis, and programming languages. Objects combine the behavior and the data of a component into conceptual and even physical entity. The interaction between objects is limited by an interface that abstracts from the particular implementation of the objects. Two key prerequisites were essential for the overwhelming success of object orientation in software design. First, object oriented methodology provides adequate solutions for major design challenges, e.g., the structure and the components of complex systems can be mapped naturally onto compositions of objects. Second, efficient tools — and here especially compilers for object oriented languages — fostered designers' productivity.

W. Müller et al. (eds.),
SystemC: Methodologies and Applications, 217–246.
© 2003 *Kluwer Academic Publishers.*

For several years now the dominant paradigm in hardware design has been register transfer level (RTL). With its typical building blocks such as register, arithmetic/logical unit, data path, and finite state machine it is already a significant abstraction from the lower levels such as transistor or gate level which were used before. From a structural perspective systems at RTL are compositions of blocks connected and synchronized via signals. Only by the productivity boost gained from the shift from gate level to RTL the opportunities offered by the rapid progress in semiconductor manufacturing could be realized in the past. For a few years there is also the possibility of behavioral level modeling, adding some further abstraction to RTL. With behavioral level modeling the designer can focus more on algorithmic design. The main automated steps of behavioral synthesis are allocation and binding of resources and the scheduling of operations to clock cycles. Although this reduces the effort for the designer necessary in order to achieve an efficient hardware implementation of a given algorithm, the expressiveness of behavioral specifications does not go far beyond RTL.

Today hardware system complexity reached a point where the current abstraction level significantly limits the productivity. The modeling and implementation of communication and synchronization scenarios in actual SoCs already generates a growing portion of the design effort. Anticipating the progress in semiconductor manufacturing in the near future we believe that RT and behavioral level design will become as inadequate for hardware design as assembler coding already is for the majority of software systems.

Similarities between past challenges of software design and those in hardware design practice provide a clear indication that object orientation could again help solving them. Object orientation can especially help to model the complex structure of modern systems and describe communication within and between hardware entities at an adequate level of abstraction. Stepwise refinement and generalization are essential concepts for efficient IP re-use in hardware design which are, in contrast to VHDL [Ashenden, 2002] or Verilog [Thomas and Moorby, 1991], well supported by object oriented features such as inheritance and polymorphism.

SystemC^TM [OSCI, 2002a, Bhasker, 2002, Grötker et al., 2002] and SystemVerilog [Accellera, 2002b] are two prominent examples for hardware design languages already trying to improve the designers productivity by higher levels of abstraction in general, and object orientation, in particular. While these concepts are extremely useful for the modeling of complex hardware systems they lack a clear synthesis semantics and thus can not be translated automatically into hardware.

In the remainder of this article we will first give a brief overview of related work. Next, we will shortly discuss the benefits of SystemC for modeling and the fundamental problems regarding refinement and synthesis of its higher level language elements. Afterwards we will present the SystemC based *SystemC*

Plus approach[1] that provides object oriented language features in combination with a clear synthesis semantics. Lastly, we will outline two synthesis strategies that allow dealing with object orientation in hardware design and whose principle applicability has already been proven.

8.2   Related Work

Of course, the idea of introducing object orientation into hardware design is not brand new, although it is still an innovative topic and a matter of research. Over the past years there has been a variety of approaches to reach this goal based on different languages. Some of these approaches try to augment object oriented programming languages by capabilities to model hardware, e.g., by providing class libraries for modeling and simulating hardware. SystemC belongs to this class. But there are further C/C++-based approaches, such as OCAPI-xl [Vanmeerbeeck et al., 2001, Vernalde et al., 1999] and also some that are based on other programming languages, like Java [Kuhn et al., 1998]. Other approaches try to augment hardware description languages by object oriented concepts and elements. The most prominent approaches in this class are based on VHDL, like SUAVE [Ashenden et al., 1997, Ashenden et al., 1998], OO-VHDL [Swamy et al., 1995], and Objective VHDL [Schumacher, 1999, Radetzki et al., 1998].

For several reasons none of these approaches has reached a breakthrough so far. First of all, the majority of the existing approaches focuses on simulation at first instance allowing to model and simulate hardware on higher levels of abstraction but neglecting hardware synthesis. Unfortunately the lack of an automated synthesis path strongly limits the benefit of high level modeling and also limits acceptance of such approaches.

Other approaches, especially those that are based on object oriented programming languages, make only use of object orientation in order to augment a language by hardware description capabilities. In this case hardware characteristics like signals, entities, and typical hardware data types are usually provided in a class library, but this is only a way of avoiding the introduction of new keywords. These approaches often do not allow one to apply object orientation for modeling system functionality itself, e.g., by providing user defined data types.

Another set of approaches strongly limits the degree of object orientation that can be applied or does not consequently realize object orientation, e.g., abandoning polymorphism. But this is also closely coupled with the abilities, or, more precisely, disabilities, of existing synthesis solutions.

---

[1] In fact, it is an additional class library like the Master–Slave Library combined with a modeling/coding guideline.

Those approaches which are based on existing hardware description languages suffer from these languages being usually not the optimal basis for the introduction of object oriented extensions and that they can only hardly be extended without severe modification of the original syntax and semantics.

The approach presented in this article picks up a variety of ideas that has already been formulated similarly in those previous works. But in contrast to them its emphasis is clearly placed on hardware synthesis from the early beginning, while trying not to seriously limit the degree of object orientation that can be applied and choosing a language which has good prospects of becoming the standard system level modeling language in the future as a basis.

8.3   High Level Hardware Modeling with SystemC

The reason why, in contrast to traditional imperative hardware languages, SystemC allows modeling at higher levels of abstraction is obviously that it is based on C++. From a historical point of view C++ only built the implementation framework for the reference SystemC simulation kernel and was not necessarily intended to be also used for modeling functionality. This is reflected in the first SystemC synthesis subsets [Synopsys, 2002a, Synopsys, 2002b], which basically only comprise a restricted C subset similar to VHDL or Verilog. But since each SystemC model can be compiled with a common C++ compiler, in principle nothing speaks against using C++ features for modeling system behavior, too. This is exactly what higher level — which means beyond RT level — modeling methodologies based on SystemC do: using C++ in order to improve the system design process itself.

8.3.1   New Perspectives

There are a lot of areas in hardware design which can greatly benefit from object oriented modeling. For instance, the C++ template mechanism combined with the ability to declare modules as templates and the ability to pass classes as template parameters, provide a great chance for modeling highly parameterizable components. Think of passing different kinds of communication interfaces, some serial transceiver, a parallel interface, or a PCI bus interface, just as template parameters, to optimally tailor a system to the actual needs. Likewise, different implementations of an arithmetical unit may be passed as a parameter, for instance, a small but slow adder in one case or a fast but big adder in another case. This is even superior to the generic mechanism in VHDL, where only scalar values can be passed as generic parameters.

There is a variety of data structures in hardware which can be excellently modeled applying object oriented techniques. A hierarchy of data packages is only one example. A common base class would capture common attributes and methods such as destination address and CRC, and the derived classes could implement specialized package formats. Applying polymorphism as well would allow one to store data packages of different formats in the same buffer, at the same time. And something like a router could be modeled in such a way that it is able to handle various different formats of data packages without knowing each format in detail, as long as the address format and its access method were the same.

Modeling new data types is another natural domain of object orientation. For instance, different number formats like fixed point and complex can be easily modeled as classes, without having to extend the basic type system. Structuring these numerical formats in a hierarchy and again applying polymorphism would allow one to model an ALU that could handle numbers of different formats without having to know about their details. The knowledge of how to perform a certain operation on a certain number format would be encapsulated in the definition of the format in itself.

Also communication could be pushed to new levels. Instead of communicating via signals and having to deal with interfaces which may consist of hundreds of ports, the use of methods could greatly improve communication modeling, since a method interface is much more readable than just a set of signals. This is also the basic idea behind the so called transaction level modeling [Grötker et al., 2002], one of the first outcomes of the new modeling capabilities provided by SystemC. Here the possibility of encapsulating even complex communication interfaces in a concise set of methods allows one to cleanly separate communication and behavior in a design. The implementation details of the communication components, namely the channels, can be hidden behind a small and handy interface. Even completely different communication channels, such as a simple buffer or some kind of complex network, could be exchanged while keeping the same interface.

Of course the greatest benefit for modeling can be expected from a sophisticated combination of the different facets of object oriented modeling, allowing one to design hardware components that are flexible in use and highly maintainable at the same time.

8.3.2   And Synthesis?

All this is already possible with SystemC, but with one significant drawback: it does not go together with hardware synthesis, at least not with the traditional behavioral and logic synthesis tools and techniques. One reason for that is that hardware synthesis from object oriented specifications is still an innovative field, lacking sufficient experience and approved techniques. Already the set of synthesis tools that support SystemC is only a small one, not even talking of any support for object orientation. What these tools, for instance, the Synopsys CoCentric®SystemC Compiler [Synopsys, 2002b, Synopsys, 2002a], usually support is a restricted subset of SystemC that is comparable to common VHDL synthesis subsets. A further reason is that features like the SystemC channels, primitive as well as hierarchical ones, unfortunately lack a clear synthesis semantics[2]. It is not clearly specified how to synthesize a channel and what hardware should be synthesized from it.

Take, for example, the primitive channel sc_fifo. It is a nice example for a data container in the object oriented sense, but its behavior would be quite unrealistic for a real world hardware implementation, mainly because it is untimed. As a model for real hardware sc_fifo just lacks adequate functional and timing accuracy, and therefore does not correctly reflect the behavior of an actual implementation. Although synthesis support for it could be integrated in principle into a tool, this would be rather an interpretation of what sc_fifo is, for instance, a dual ported register file, than a direct synthesis from its description. And it would be a proprietary solution, since other synthesis tools may map it to another kind of hardware, preventing designers from implementing own synthesizable channels.

It is a general problem of the channels in SystemC v2.0 that the way to implement channels conceptually prevents, or at least complicates a direct synthesis from their specification. Refinement of a channel down to a cycle accurate level usually requires some information about the way a channel works and how it should be refined. This information can not be derived directly from its implementation, making an automation of the refinement hardly possible. One reason for that might be that SystemC as a system level description language does not necessarily address only hardware, and that channels in general, and sc_fifo, in particular, represent even more abstract communication mechanisms. But SystemC is also clearly a hardware description language, and the reasoning above will not really satisfy a hardware designer who makes use of channels in his design and who would like to combine the modeling power of channels with automated hardware synthesis.

In summary, SystemC builds a very good starting point for combining hardware design with object orientation but also reveals some significant drawbacks regarding hardware synthesis starting from higher levels of abstraction. In the following we will present a possible approach to how the level of abstraction for hardware modeling, based on SystemC and in combination with new synthesis techniques, can be raised without having to abandon automatic synthesis.

8.4   The SystemC Plus Approach

The *SystemC Plus approach* [Grimpe et al., 2002] basically comprises an additional class library, like the Master–Slave Library [OSCI, 2002b] available

---

[2]sc_signal is excluded from this consideration, since it is a basic primitive for hardware modeling with a well defined synthesis semantics rather than just another channel.

for SystemC, and modeling/coding guidelines [Ashenden et al., 2001], together defining an extended SystemC/C++ synthesis subset. Its goal is to provide object oriented hardware modeling in SystemC, combined with a clear synthesis semantics and a hardware-like simulation semantics for the higher level language constructs. For this purpose, on the one hand the approach specifies restrictions and coding guidelines on the use of C++ and also some on SystemC, but on the other hand it allows the use of most of the C++ object oriented language features for modeling or provides alternative constructs in its class library that better fit hardware synthesis.

A clear synthesis semantics means that the SystemC Plus synthesis subset includes only those higher level language constructs, for which a mapping to behavioral equivalent constructs on RT and behavioral level is clearly specified. Analogously to the term *behavioral synthesis*, we will refer to this mapping as *object oriented synthesis*, or *oo synthesis* for short, in the following.

A hardware-like simulation semantics means that the simulation behavior of the higher level language constructs in the synthesis subset correctly reflects the behavior of a real world hardware implementation, in terms of functionality and cycle accurate timing. Note that this equivalence only holds for specified behavior, meaning that, for instance, the initial value of an uninitialized variable or data member may differ before and after synthesis.

The SystemC Plus synthesis subset is based on a common SystemC behavioral/RT level synthesis subset[3], which, apart from the basic elements that are necessary for describing hardware — modules, processes, signals, ports, and data types typical for hardware design — provides only a restricted subset of C for modeling. This subset is therefore widely semantically equivalent to the VHDL behavioral/RT level synthesis subset[3], and can be processed with traditional synthesis techniques and tools. The SystemC Plus approach augments this subset by the full C++ class concept for object oriented modeling, a polymorphism concept that is not based on pointers, and an alternative concept for SystemC channels, that better fits hardware synthesis needs. Features excluded from the subset include, for example, file I/O, static data members, global variables, events, channels (except for sc_signal), and dynamic memory allocation and deallocation (new and delete). Most of these restrictions are not directly related to object oriented modeling and are therefore not addressed by the oo synthesis techniques which are illustrated later on. They are either of a principle nature, like file I/O, or they are in consequence of limitations of the behavioral and logic synthesis which follows oo synthesis (refer also to Figure 8.2 in Section 4) and do not overlap with object orientation.

---

[3]Since this is not fully standardized, this basically means the subset that is most common supported by existing tools.

The basic framework for using object orientation in hardware models on a level similar to C++ is provided by the SystemC Plus approach through support of the following C++ class related features:

- Class declarations and class instances (objects);

- Inheritance, including multiple inheritance, and virtual base classes;

- Declaration and use of member functions, including overloading, and constructors;

- Operator overloading;

- Virtual function declaration (refer also to the following discussion of polymorphism);

- Class templates, including scalar and type parameters.

One of the most obvious restrictions on the use of C++ is, of course, the ban on dynamic memory allocation and deallocation, and in combination with it the ban on using pointers. From a software programmer's point of view this may give the impression of a serious limitation, but from a hardware designer's point of view this makes sense. Hardware, in contrast to software, is of a static nature, which means that hardware resources can not just be created and destroyed during runtime[4], whilst allocating and freeing memory in software is not a matter within the given physical limits of the available memory. Pointers are banned for two reasons. First, in the absence of dynamic memory management the benefit of pointers is very limited, and using them makes only little sense. And second, although possible, pointer synthesis is complex and tends toward producing inefficient hardware. Simply speaking, the cost–value ratio does not speak for pointer synthesis.

8.4.1 Polymorphism

In the absence of pointers the question arises of how polymorphism, which is, of course, one of the most powerful object oriented features, can be realized, since this mechanism is completely based on pointers in C++. In other programming languages, like Ada95 [Burns and Wellings, 1998, Barnes, 1998], the support for polymorphism does not necessarily depend on pointers nor even references. An alternative realization is, for example, by means of *tagged objects*. A tagged object can dynamically change its type, or, more precisely, its class type, during run time. The information of to which class such an object

---

[4]This holds at least for traditional ASIC design. Things look different for reprogrammable chips, but for this kind of technology the borders of hardware and software are blurring anyway. However, it is arguable whether reprogrammable logic will eventually completely replace hard wired logic in the foreseeable future.

actually belongs is *tagged* to the object. Based on this information dynamic dispatching of methods can be realized. The major difference compared with the native C++ polymorphism concept and what makes tagged objects better fit synthesis is that a tagged object physically exists, that means provides its own state space, making the need for dynamic resolution of object references dispensable.

For this reason polymorphism in the SystemC Plus approach is based on tagged objects, called *polymorphic objects* there, with basically the same behavior and assignment rules as used in C++. In contrast to C++ these polymorphic objects are self-contained objects as discussed above, and not just references to objects. Since the really interesting feature of polymorphism is dynamic dispatching of member functions, it is naturally supported by SystemC Plus' polymorphic objects. Each member function of a polymorphic object's root class that is declared virtual is dynamically dispatched during runtime. That means the concrete function implementation that is executed on a call depends on the actual type of a polymorphic object.

In the following the usage of this polymorphism mechanism is illustrated and compared to the native C++ polymorphism mechanism. First, a simple class hierarchy is implemented. A typical example for a class hierarchy with relevance to hardware design is a hierarchy of instructions, e.g., arithmetical operations for an ALU:

```
class ALUInstruction {
public:
  POLYMORPHIC( ALUInstruction ) // Class is now "tagged"

  ... // constructor, other member functions, ...

  virtual
  int
  execute(); // Does nothing

private:
  int m_leftOperand;
  int m_rightOperand;
};

class Mult :  public ALUInstruction {
public:
  POLYMORPHIC( Mult ) // Class is now "tagged"

  ... // constructor, other member functions, ...
```

```
  virtual
  int
  execute(); // Implements multiplication
};

class Add :  public ALUInstruction {
public:
  POLYMORPHIC( Add ) // Class is now "tagged"

  ... // constructor, other member functions, ...

  virtual
  int
  execute(); // Implements addition
};
```

The newly introduced specifier POLYMORPHIC is used in order to tag a class type. Only such tagged class types can be used as root classes for polymorphic objects, and consequently only instances of tagged classes can be assigned to polymorphic objects. But including the POLYMORPHIC specifier in a class declaration does not limit its use, thus it can still be used as any other ordinary class. Owing to declaring the function execute() virtual, it will be dynamically dispatched when called on a polymorphic object.

Next, basically the same code sequence is presented two times, first realized with pure C++ and afterwards based on polymorphic objects:

```
// C++ as usual:
ALUInstruction *polyInstruction;
Mult multiplication( 2, 5 );
Add addition( 2, 5 );
int result;

polyInstruction = &multiplication;
result = polyInstruction->execute(); // multiplies 2 with 5
polyInstruction = &addition;
result = polyInstruction->execute(); // adds 2 to 5

// The same with polymorphic objects:
PolyObjetc< ALUInstruction > polyInstruction; // Declares a
        // polymorphic object with root class ALUInstruction
Mult multiplication( 2, 5 );
```

```
Add addition( 2, 5 );
int result;

polyInstruction = multiplication;
result = polyInstruction->execute(); // multiplies 2 with 5
polyInstruction = addition;
result = polyInstruction->execute();  // adds 2 to 5
```

In both cases the execute() operation is dynamically dispatched and the implementation being invoked depends on the type of the last object being assigned. So far both solutions look very similar and also behave very similarly in this example. But if operations that change the state of an object were invoked the differences between both solutions would become apparent.

A pointer only references an object, and all operations that are invoked on the pointer take direct effect on the referenced object. If, for instance, the value of one of the operands is set via the pointer polyInstruction directly after the address of object multiplication has been assigned to it, the operand of object multiplication is modified.

A polymorphic object is a self-contained object, not a reference, and will hold a real copy of an object in an assignment. All operations invoked on a polymorphic object only take effect on the copied values and not on any assigned object. Another consequence of this difference is that a root class of a polymorphic object can not be an abstract class, since memory is already allocated during its instantiation. However, the clear advantage of a tagged object approach is that it provides the expressive power of polymorphism but fits hardware synthesis much better, since it avoids expensive lookup operations for resolving pointers. How synthesis is done in detail is illustrated later. It shall be understood that polymorphic objects behave functionally accurately with respect to an actual hardware implementation.

8.4.2   An Alternative for Channels

As mentioned earlier, SystemC channels lack a clear synthesis semantics and reveal some significant drawbacks regarding hardware synthesis. They are therefore excluded from the SystemC Plus synthesis subset. On the other hand, channels undoubtedly provide some nice features for modeling. In particular, they allow one to model inter-process communication in a comfortable way, on a much higher level of abstraction than that based on signals. And this is indeed something that can significantly improve the design process. That is why the SystemC Plus approach provides an alternative concept for inter-process communication, not being just as powerful and flexible as channels, but still on a comparable level of abstraction and reflecting the behavior of hardware and synthesis needs much better. It is based on so called *global objects*.

A global object is, in principle, an object that is declared as a shared variable, as member of an sc_module, visible and accessible by all processes that are declared within the same module. Before going into more detail let us first have a quick glance at the most distinct features which come along with a global object:

- Mutual exclusive access; only one client is granted access at a particular time;

- Different scheduling strategies; the scheduling strategy that is used for solving concurrent accesses can be chosen by the user. Scheduling is automatically invoked;

- Arbitrary number of clients; an arbitrary number of client processes can access each global object. The number of clients does not have to be specified explicitly;

- Hardware-like simulation semantics; timing (cycle accurate) and functionality of a real world hardware implementation are accurately reflected by a global object. It behaves like a synchronous component;

- Clear synthesis semantics; it is clearly specified to what kind of hardware a global object is mapped. Synthesis can be fully automated;

- Guarded functions; function execution depends on the result of the evaluation of an associated Boolean expression, called the guard or the guard condition;

- Atomic guard evaluation; evaluating the guard condition and, dependent on the evaluation result, function execution is an atomic action;

- Binding; like ports, global objects of the same type can be bound to each other, thus enabling inter-module communication.

A global object is a class template which requires two parameters; the first parameter must be a scheduler class, which is used in arbitrating concurrent accesses, the second parameter is any user-defined class, implementing the intrinsic behavior of the object. An exemplary declaration looks as follows:

```
SC_MODULE( ProdCons ) {

    // Global object with scheduler "RoundRobin" and user-
    // defined class "FIFO< class Type, SizeType Size >":
    GlobalObject< RoundRobin, FIFO< int, 8 > > sharedBuffer;

    ... // other declarations and module constructor
};
```

The SystemC Plus class library already includes a few pre-defined schedulers, but designers can also define their own scheduling classes, following some coding guidelines. The central element of a scheduler is the `schedule()` function, which must be accordingly overloaded by each user defined scheduler class. The `schedule()` function is automatically invoked each time a service is externally requested on a global object in order to determine the next request that will be granted. Basically it is a normal synthesizable member function of the scheduler class from which the scheduler is actually synthesized, without any additional information necessary to be specified by the user.

The intrinsic functionality of a global object is given by the class that is passed as second parameter. In the above example it is a FIFO buffer, a class template itself, that can store up to eight elements of type integer. Although the second parameter can be any ordinary user defined class, one additional coding guideline has to be taken into consideration: each of its member functions which is called by a client process, must be declared as a so called *guarded method*. Taking the class declaration of the FIFO class that is used above, this looks as follows:

```
typedef unsigned int SizeType
template<class Type, SizeType Size>
class FIFO { public:
  FIFO();

  GUARDED_METHOD( Type,           // return type
                  read(),         // signature
                  !isEmpty() );   // guard condition

  GUARDED_METHOD( void,
                  write( const Type & element ),
                  !isFull() );

  GUARDED_METHOD( bool,
                  isFull() const,
                  true );

  GUARDED_METHOD( bool,
                  isEmpty() const,
                  true );

protected:
  Type m_buffer[Size];
  bool m_empty;
```

```
bool m_full;
SizeType m_top;
SizeType m_bottom;
SizeType m_numberOfEntries;
};
```

A guarded method declaration associates a *guard condition* — a Boolean expression — with a member function. This guard condition is evaluated during scheduling and only those clients may be granted which are calling methods with a true guard condition. Evaluation of the guard condition and the execution of a guarded method are an atomic action. No other method execution can interfere, making the usage of an additional semaphore obsolete. Taking the above FIFO as an example, it only makes sense to allow execution of the `write()` method if there is storage space left in the buffer, thus the guard condition of this method checks whether the FIFO is not already full. Likewise, the guard condition of the `read()` method checks whether there is an element left that can be read out. 'true' as the guard condition means that a guarded method is always ready for execution. To specify 'false' as the guard condition means that the associated guarded method can never be externally invoked, which usually would not make sense. It is also possible to test parameters that are passed to a guarded method. An extended `write()` function may, for instance, declare a parameter that specifies the number of elements which should be put into the FIFO. An associated guard condition could check whether the buffer has enough empty slots left for this action.

Adding guarded method declarations to a class does not prevent its local use. Every class, such as FIFO, declaring any guarded method can be instantiated and used like any other ordinary class, as a local instance within functions and processes. In this case the guard mechanism does not take effect.

Clients of a global object have to be synchronous processes, namely SC_CTHREADs. Each client process which wants to access a global object first has to *subscribe* to it. Subscribing to a global object is done by means of a *global object subscription statement* that must be located in the reset part of a client process. For subscription an unsigned integer number can be optionally specified as an argument which can be used by the scheduler, for instance, as a priority. Once a client process has subscribed to a global object it may include any number of method calls to this global object, so called *global calls*, in its main body:

```
void producer { // declared as an SC_CTHREAD somewhere
  int val;
  if ( reset == true ) {
    val = 0;
```

```
        sharedBuffer.subscribe(); // global object
                                  // subscription statement
      }
      wait();
      while( true ) {
        GLOBAL_PROCEDURE_CALL( sharedBuffer, put( val ) );
        val += 1;
      }
    }
```

Global calls are always blocking. That means a client is blocked at the call site until its request is finally granted. Note that the global object based communication mechanism in the SystemC Plus approach can not guarantee freedom from for deadlocks and starvation.

Though a global call looks a bit more 'ugly' and may even feel a bit less object oriented than a usual SystemC channel access owing to the usage of a macro, it is still method based and easy to use whilst providing all the features that have been listed previously. And, as another advantage, the temporal behavior of an actual hardware implementation is accurately reflected in simulation also taking the necessary communication between a client and a global object into account. That means that each global call, even if immediately granted, will block a client for some clock cycles, since handling the communication and the execution of the called method naturally consumes some time in real hardware. Thus the user receives an early but realistic impression of the temporal behavior of the modeled hardware even before performing synthesis, while keeping all the benefits of high level modeling.

In order not to limit communication via global objects only to processes being located in the same module, global objects, of the same type, which means instantiated with the same template arguments, can be bound to each other like ports:

```
SC_MODULE( ProdCons ) {
  GlobalObject< RoundRobin, FIFO< int, 8 > > sharedBuffer;

  Producer * producer1; // sub-module with global object
                        // "sharedBufferProd"
  Consumer * consumer1; // sub-module with global object
                        // "sharedBufferCons"

  ...  // other declarations
  SC_CTOR( ProdCons ) {
```

232

```
producer1 = new Producer( "producer1" );
producer1->sharedBufferProd( sharedBuffer );
... // port bindings

consumer1 = new Consumer( "consumer1" );
consumer1->sharedBufferCons( sharedBuffer );
... // port bindings
    }
};
```

By this mechanism processes can communicate with each other throughout the whole design hierarchy. The bound global objects virtually behave like a single object.

While a single global object is somehow similar to a SystemC primitive channel, it is also possible to model something similar to SystemC's hierarchical channels with global objects, using the binding mechanism. For achieving this goal global objects are just used as an interface to a module. Different modules are connected with each other by binding the global objects which form their interfaces to each other. The 'Channel' in Figure 8.1 connects 'Block1' and 'Block2' in the same way a hierarchical channel in SystemC would. Each of the components in this topology can be either a simple module or a complex hierarchical design. Even if the 'Channel' module implements an arbitrary complex kind of communication, 'Block1' and 'Block2' can still communicate with each other by simple method calls, since the connection is established via global objects.

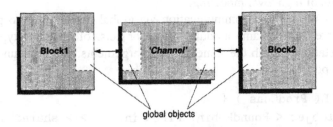

Figure 8.1.   Using global objects for 'emulating' hierarchical channels

Thanks to the fixed access mechanism for global objects, formed by the communication protocol, scheduling and guard mechanism, and its separation from the user defined behavior, the behavior itself can be easily extended and modified by the user without having to track the synthesis semantics each time.

To summarise, global objects have some clear advantages over SystemC channels for hardware modeling and synthesis. A global object accurately reflects the behavior of the hardware that is synthesized from it, thanks to built in mechanisms which cannot be circumvented by the user. Global objects possess a scheduling mechanism which guarantees mutual exclusive access. The integrated guard mechanism makes the use of additional semaphores or monitors obsolete. A global object can have an arbitrary number of clients, without having to explicitly specify that number, and global objects can be bound to each other, thus enabling communication throughout a whole design hierarchy. All this is provided in a way that clearly separates between the user defined behavior of a global object that is directly synthesized and its basic functionality that rather serves as a framework for the behavior. Most of these points could also be achieved with SystemC channels, but only completely handcrafted and without any special conceptual support. And, in particular, the last mentioned point regarding synthesis does not seem to be realizable with channels as provided in SystemC v2.0.

8.5   Hardware Synthesis from Object Oriented Models

One characteristic of a hardware description language is that it typically serves as the starting point for automated hardware synthesis. Introducing a new level of abstraction, or, more precisely, object orientation, into hardware design to improve the design process therefore makes only much sense if the synthesis flow does not get stuck at this level. Manual refinement is error-prone and time consuming and does not seem to be a convenient solution for the ever increasing system complexity. Hence a clear synthesis semantics, that means a mapping to lower levels of abstraction, is required for object oriented language constructs that can serve as basis for automated synthesis. In the following the basics of even two such mappings are outlined — object splitting and a mapping to bit vectors —, each of them having some advantages and disadvantages in specific situations. In particular, the latter is strongly inspired by [Radetzki, 2000]

Figure 8.2 illustrates the synthesis flow starting from object oriented specifications. Object oriented synthesis is settled on top of existing and approved synthesis techniques, and complements them but does not replace them. The figure also clarifies that object orientation is a modeling paradigm that is orthogonal to behavioral and RT level modeling, it can be applied at both levels. But it would offer the greatest modeling benefit in conjunction with effective behavioral synthesis.

8.5.1   Synthesizing Objects

As already indicated by the term *object orientation* the main characteristic of an object oriented language is the use of objects as a basic language construct. But this kind of data structure is not, or only very limitedly, supported by

234

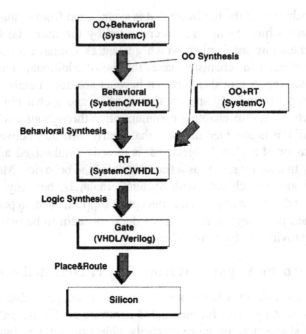

Figure 8.2. Synthesis flow starting from object oriented hardware specifications

behavioral and RT level synthesis tools. Hence the first and most important step in object oriented synthesis is to eliminate, or, more precisely, to replace, all objects. As alluded to previously, we will present two approaches that lead to this goal.

The first approach is to split objects into individual variables. Leaving member functions aside at first — they will be discussed in the next section — an object can be simply regarded as a set of variables, namely its data members. Thus an object declaration can be replaced by a set of individual variable declarations, each of them representing one of the original data members. Access to data members is replaced by access to the extracted variables. Assignments from and to objects are transformed into a sequence of individual assignments from and to each variable that has been extracted from the original objects. Signals and ports of class type are split and treated in the same way. The access specifiers public, protected, and private, although being useful for modeling, obviously have no meaning for synthesis[5], and are therefore simply ignored.

---

[5]Here we presume that the design being synthesized is semantically and syntactically correct, i.e., can be successfully compiled with an ANSI compliant C++ compiler.

Inheritance, even multiple inheritance, does not pose a special problem for synthesis. Simplifyingly speaking inheritance represents a kind of compiler supported sophisticated copy&paste, and the same effect can be achieved by copying all data members that are inherited by a class directly into the inheriting class. This is performed before an object is split, thus ensuring that each class instance will include a full set of data members, also including all inherited ones.

Take, for example, the following simple piece of code:

```
class Base {
public:
    bool m_flag;
};

class Derived :  public Base {
public:
  int m_number;
  char m_character;
};

Derived obj;
int variable;

obj.m_flag = true;
obj.m_number = 42;
variable = obj.m_number + 1;
obj.m_character = 'z';
```

After splitting the used object the code would look as follows:

```
bool obj_m_flag;
int obj_m_number;
char obj_m_character;
int variable;

obj_m_flag = true;
obj_m_number = 42;
variable = obj_m_number + 1;
obj_m_character = 'z';
```

236

By iteratively repeating object splitting nested objects, i.e., objects that themselves contain objects as members, also are completely flattened. If an object contains arrays[6] as members, these members are preserved as arrays, and are not further split into their single elements. But each multi-dimensional array is mapped to a one-dimensional array during synthesis. Preserving arrays and mapping them to one dimension keeps the possibility of mapping arrays to memories in later synthesis steps. If the elements of an array are objects themselves, they are mapped to bit vectors, as presented in the following. At the end of the complete process, only variables of scalar type and arrays with elements of scalar type are left, which should be processable by any behavioral or RT synthesis tool.

The second approach for replacing objects is to map them to bit vectors, which means that each object is represented by a single bit vector. One motivation for this kind of mapping is that object splitting is not always a straightforward process, especially not where objects are treated on the whole, for instance, as function arguments or return values (refer also to the following section). But the main motivation is that this kind of mapping allows one to preserve arrays of objects, which means arrays with class instances as elements, instead of splitting them into hundreds or even thousands of individual variables. This is, in particular, useful for allowing an efficient mapping of arrays to memories, since each element could be completely mapped to exactly one directly addressable memory entry.

In order to map an object to a bit vector, first, the bit width of each data member is determined, for example, 32 bits for a member of type integer, 1 bit for a Boolean member. The sum of the individual bit widths will give the size of the bit vector that replaces the complete object. The original data members are mapped onto slices of this bit-vector as illustrated in Figure 8.3

Coming back to the previous example we now transform the used object into a bit vector:

```
sc_biguint< 41 > obj; // 1(bool) + 32(int) + 8(char)
sc_bigint< 32 > variable;

obj.range( 0, 0 ) = 1;
obj.range( 32, 1 ) = 42;
variable = ( sc_bigint< 32 > )( obj.range( 32, 1 ) ) + 1;
obj.range( 40, 33 ) = 122; // 122 = ASCII 'z'
```

---

[6]Arrays of bits and logic values, namely, bit vectors and logic vectors, are excluded from this consideration, since here they are regarded as their own data types comparable to scalar data types.

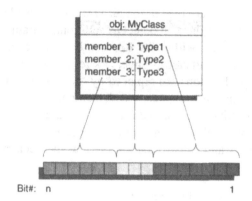

Figure 8.3.   Transforming an object into a bit vector

In fact, an sc_biguint is used to represent a synthesized object because it has several advantages. It behaves and can be used like a bit vector. Its size can be arbitrarily chosen allowing one to represent even large objects. And most importantly, an sc_biguint still allows the performing of arithmetical operations on its slices and the mixing of its slices in expressions with other variables and literals of C integer types. Unfortunately the transformation into a flat type-less bit vector may cause some members to lose their original type, like signed integer or Boolean members. But, as demonstrated in the above example, this information can be restored by appropriate type casting where necessary.

The two different synthesis strategies presented here are not mutually exclusive, which means they can be applied together on the same model. In fact, only a combination of both mappings would lead to the most efficient synthesis result in many cases. As already alluded to above, it makes sense, for example, to split an object which itself contains an array of objects, but to preserve the array and to map its elements to bit vectors in order to allow memory mapping of the array.

8.5.2   Member Functions

Usually not only data members but also member functions belong to an object, forming its interface to the environment. When objects are replaced in the manner demonstrated above, necessarily also all member function declarations and member function calls must be replaced somehow.

The most important difference between a member function and a non-member function is that the object on which it is invoked forms an additional scope for the member function, which means that all members declared in that scope are visible and accessible for it. In the course of replacing objects this implicitly given scope is destroyed and must be restored somehow. This is achieved by either passing the set of variables or the bit vector — dependent on the chosen object mapping — that replace an object as additional parameters to a member function. Member access within the function body is replaced by access to the respective formal parameter or slice. As result of this transformation all member function declarations are replaced by 'normal' non-member function declarations in the global name space, and all member function calls are replaced by calls to the respective transformed function declaration.

For illustration we will first augment a previous example by some member function declarations and calls to them (note that the declaration of the class Base is left out):

```
class Derived :  public Base {
public:
  Derived() :  m_number( 0 ),
               m_character( 'a' ) {
  }

  void
  incNumber( const int val ) {
    m_number = m_number + val;
  }

  int m_number;
  char m_character;
};

Derived obj;  // Invokes default constructor
obj.incNumber( 10 );
```

The transformation of this code in case of object splitting would give the following result:

```
void
_Derived_incNumber( bool & _this_m_flag,
                    int & _this_m_number,
                    char & _this_m_character,
                    const int val ) {
  _this_m_number = _this_m_number + val;
}
void
```

```
_Derived_Derived( bool & _this_m_flag,
                  int & _this_m_number,
                  char & _this_m_character ) {
  _Base_Base( _this_m_flag );   // Default constructor of
                                // Base is invoked
  _this_m_number = 0;
  _this_m_character = 97; // 97 = ASCII 'a'
}

bool obj_m_flag;
int obj_m_number;
char obj_m_character;

_Derived_Derived( obj_m_flag, obj_m_number, obj_m_character);
_Derived_incNumber( obj_m_flag, obj_m_number,
                    obj_m_character, 10 );
```

Apparently there is some potential for optimization. The parameters _this_m_flag and _this_m_character are not used within the body of the function, and thus could be omitted. Such dead parameters, as well as dead variables, can be identified and eliminated with classical analysis and optimization techniques, for instance, data-/control-flow analysis after all object oriented elements have been completely removed from a model.

If objects are replaced by bit vectors the previous example looks as follows:

```
void
_Derived_incNumber( sc_biguint< 41 > & _this_,
                    const int val ) {
  _this_.range( 32, 1 ) =
    ( sc_bigint< 32 > )( _this_.range( 32, 1 ) ) + val;
}

void _Derived_Derived( sc_biguint< 41 > & _this_ ) {
  _Base_Base( _this_.range( 0, 0 ) );
  _this_.range( 32, 1 ) = 0;
  _this_.range( 40, 33 ) = 97;
}

sc_biguint< 41 > obj;
```

```
_Derived_Derived( obj );
_Derived_incNumber( obj, 10 );
```

This time, applying the same optimizations for eliminating dead members would lead to a bit vector of reduced size. But what makes optimization more complex in this case is that for different functions the same object may be represented by bit vectors of varying size, dependent on the set of members that is actually used by a function. Additionally, the absolute position of a slice representing the same member may vary. For instance, if one function uses the members m_flag and m_number and another one only uses the member m_number the start index of the slice representing m_number would be 1 in the first case and 0 in the latter case (presumed that optimizations are applied).

8.5.3   Class Types as Formal Parameter and Return Types

Handling formal parameters of class type for synthesis is quite simple: they are either replaced by a set of formal parameters representing a split object or their type is transformed into a bit vector of appropriate size, again dependent on the chosen synthesis strategy for objects. Access to members of such parameters is replaced by access to the respective formal parameter or slice, as already illustrated in the previous section.

When the return type of a function is a class type things remain relatively easy, as long as objects are mapped to bit vectors. In this case the return value will still stay a single value; in principle only its type changes. But since the type of all targets to which the returned value of the function is assigned changes in exactly the same way, basically nothing more has to be taken into consideration for synthesis.

In the case in which an object is split synthesis is more complex, because a function with a class type as return type would no longer return just a single value but a set of separate values[7]. Since this is obviously not possible, a function must first be transformed into a procedure — a function with void return type — with an additional formal parameter that is used to store the return value of the original function. Next, all call sites of the function must be replaced by a temporary variable that is initialized by means of a call to the transformed procedure, directly preceding the call site. After all functions with class types as return type have been replaced this way, object splitting can be applied as usual. The following sample code illustrates this process:

```
class Point { ... };
```

---

[7]It is presumed that the split object has more than one data member.

```
class Line {
  ...
  Point
  getStartPoint() const {
    return( m_startPoint );
  }

private:
  Point m_startPoint;
  Point m_endPoint;

};

Point p1( 1, 1 );
Point p2( 3, 3 );
Line l1( p1, p2 );
Point dummy;

dummy = l1.getStartPoint();
```

And after synthesis:

```
void _Line_getStartPoint( int & _return_xPos,
                          int & _return_yPos,
                          int & _this_m_startPoint_xPos,
                          int & _this_m_startPoint_yPos,
                          int & _this_m_endPoint_xPos,
                          int & _this_m_endPoint_yPos ) {
  _return_m_xPos = _this_m_startPoint_xPos;
  _return_m_yPos = _this_m_startPoint_yPos;
}
...   // Declaration of split objects p1, p2, l1
int dummy_xPos;
int dummy_yPos;
int _temp0_xPos;
int _temp0_yPos;

_Line_getStartPoint( _temp0_xPos, _temp0_yPos,
                     l1_startPoint_xPos, l1_startPoint_yPos,
                     l1_endPoint_xPos, l1_endPoint_yPos );
```

```
dummy_xPos = _temp0_xPos;
dummy_yPos = _temp0_yPos;
```

This example could be further optimized by omitting the temporary variables and directly passing dummy_yPos and dummy_xPos as parameters to the function. But this is only a simple case and would not be possible, for example, if a function call was used as an argument for another function call, which means if dealing with anonymous objects.

8.5.4   Polymorphism

A polymorphic object is different from a normal object, because it must be able to store objects of different class types, which means with different sets of members, and its type must be dynamically determinable in order to realize dynamic dispatching of methods. The latter requirement is achieved by adding an 'artificial' data member — the so called *tag* — to a polymorphic object for synthesis. The tag enumerates the different possible class types of a polymorphic object and its actual value determines the actual type of a polymorphic object. Each class type of which instances are assigned to a polymorphic object is represented by a unique value, and each time an object is assigned, the tag is set to the value that represents the class type of the source object. If the source of the assignment is a polymorphic object itself, the tag value is simply copied.

By means of the tag the realisation of dynamic dispatching in a way that can be processed by behavioral and logic synthesis is no longer a big problem. It can, for instance, be handled by a switch statement with the tag as condition. Taking again the small example of polymorphism which was used in section 3, each dynamically dispatched call of the function execute() is basically replaced by the following code fragment during synthesis (the tag value is represented by an enumeration literal):

```
switch( polyInstruction_tag )
{
   case mult:
      ... // Mult::execute() is invoked
   case add:
      ... // Add::execute() is invoked
   default:
      ... // ALUInstruction::execute() is invoked
}
```

Satisfying the second requirement, making a polymorphic object big enough to store any object ever being assigned, can be relatively easily achieved if objects are mapped to bit vectors. The size of the bit vector that represents a polymorphic object must be chosen only as large, as the size of the largest bit vector representing any assigned object plus the size needed for the tag $(= \lceil \log_2(|AssignedClassTypes|) \rceil$ in binary encoding). The tag is located at a fixed position in the bit vector, whilst assigned objects, dependent on their size, may only occupy a part of it.

If polymorphic objects are split the maximum set of variable declarations that is necessary to replace a polymorphic object is formed by the disjunctive union of all data members of all classes of which instances are assigned to the polymorphic object. But the size of this set is only an upper bound, since instances of classes which do not have a direct inheritance relationship, for example, siblings, can share further variables. Compared to mapping objects to bit vectors, object splitting has the disadvantage that dealing with varying sets of variables for the same object and, in particular, 'variable sharing' makes the synthesis process more complex. Furthermore, determining the bit vector as described above automatically ensures that exactly the least necessary storage space is allocated, which is much more complex and not always possible for object splitting.

8.5.5   Global Objects

In order to make a global object processable by behavioral and logic synthesis, it is basically transformed into the client/server structure which is illustrated in Figure 8.4. The global object itself is transformed into a set of synchronous and asynchronous processes implementing the scheduling/arbitration, the guard evaluation, and the services (methods) that are provided by a global object. The method based communication between client processes and the global object is replaced by a signal based communication via so called *channels*[8], using a synchronous handshake protocol for requesting and invoking services. After synthesis an external method request on a global object — a global method call — in principle is the same as a remote method invocation.

A channel consists of buses for the input and the output parameters for the method that is called, a request bus whose actual value determines which method shall be invoked (each method that is provided by a global object is identified by a unique ID for synthesis), and the necessary handshake signals.

---

[8]This must not be mixed up with SystemC channels. In this case 'channel' is just the denomination for a certain set of signals which is used for communication.

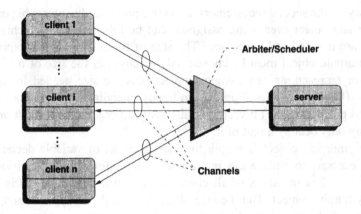

Figure 8.4. Client/server architecture being synthesized from a global object

In order to completely handle a global method call a client process executes the following steps: first, it puts the method ID onto the request bus and all necessary input parameters onto the input bus; next, it executes wait cycles until the addressed global object signals that the execution of the requested method has been started, and sets its request back. In the last step the client process again executes wait cycles until the global object signals the completion of the requested method, takes the output parameters, if any, from the output bus, and continues its regular work.

Performing this transformation for the global method call put() from the example on global method calls in section 3

```
GLOBAL_PROCEDURE_CALL( sharedBuffer, put( val ) );
```

gives the following:

```
sharedBuffer_request.write( 1 );  // ID of method "put"
sharedBuffer_parameters.write( val );  // input parameters
// wait on start of remote method execution:
wait();
while ( true ) {
  if ( sharedBuffer_done.read() != 1 ) {
    sharedBuffer_request.write( 0 );  // ID for "no
                                      // request"
    break;
  }
  wait();
}
// wait on termination of remote method execution
```

```
wait();
while ( true ) {
  if ( sharedBuffer_done.read() != 0 ) {
    break;
  }
  wait();
}
```

The last waiting for termination of the remote method execution seems to be redundant, since no output parameters are returned. But it is necessary to ensure that the one-signal handshake is correctly completed if the method execution takes more than one cycle.

8.5.6   Other Object Oriented Features

Because of limited space the synthesis techniques for a lot of other features closely coupled with object orientation, such as constructors, overloaded operators, class templates, etc., could not be presented in this article. But most of them can be treated and mapped to appropriate behavioral and RT level representations with the basic techniques illustrated above. For instance, handling overloaded operators is obviously the same as handling any other function. And a constructor is just a certain way of determining the initial values of data members and can be replaced by a normal function doing the same job.

8.6   Conclusions and Outlook

Object orientation is a powerful modeling paradigm which also offers a lot of potential for improving the hardware design process. We are confident that object orientation is one building block which could help coping with the growing design gap. As presented, applying object oriented modeling does not mean the abdication of automatic hardware synthesis, since object oriented elements can be mapped to lower levels of abstraction which can be handled with traditional and approved synthesis techniques.

Owing to its C++ roots SystemC builds an excellent framework for combining hardware modeling and object orientation. Unfortunately not all of SystemC's high level language constructs already fit hardware synthesis requirements very well. This is understandable as a tribute to system level modeling. But one advantage of SystemC is that it can be extended easily without having to touch the core language, again owing to its C++ roots. As demonstrated, this can be exploited to provide higher level language elements in a way that better reflects hardware modeling requirements.

Object oriented hardware synthesis is not just as easy as that — in particular, not efficient hardware synthesis — and it is still a research topic in its beginnings.

Whatever has been presented in this article is not the endpoint and still holds a lot of potential for improvements. And there are, of course, various critical points which have to be taken into account but could not be addressed in this short article. One is, for example, the resource efficiency, in terms of area and timing, of the hardware that has been synthesized from object oriented specifications. But there is a lot of optimization potential, and in most cases it should be possible to reduce to a minimum, or even completely eliminate, any additional overhead that may be introduced by object orientation.

Chapter 9

Embedded Software Generation from SystemC for Platform Based Design

Fernando Herrera, Víctor Fernández, Pablo Sánchez, Eugenio Villar
University of Cantabria, Santander, Spain

Abstract   The current trend in embedded system design is towards an increasing percentage of the embedded SW development cost of the total embedded system design costs. There is a clear need of reducing SW generation cost while maintaining reliability and design quality. SystemC represents a step forward in ensuring these goals. In this chapter, the application of SystemC to embedded SW generation is discussed. The state of art of the existing techniques for SW generation is analyzed and their advantages and drawbacks presented. In addition, methods for systematic embedded software generation which reduce the software generation cost in a platform based HW/SW co-design methodology for embedded systems based on SystemC is presented. SystemC supports a single-source approach, that is, the use of the same code for system level specification and verification, and, after HW/SW partitioning, for HW/SW co-simulation and embedded SW generation.

9.1   Introduction

Since the 1970s microprocessors have become common devices in electronic systems. In fact, specific processors, microcontrollers, and DSPs have been developed for this market. In these classical embedded systems, functionality is shifted to software [Lee, 2000]; that is, they include basic hardware (microcontroller/DSP, RAM, FLASH, peripherals, power unit, sensors and actuators) and nearly all the functionality is implemented in software. Existing general purpose software design techniques are not suitable because embedded software has a strong interaction with the environment, with very limited resources, strong constraints (cost, timing, power, etc...), and often is developed by engineers with little background in computer sciences [Lee, 2000]. During the 1970s and 1980s embedded software was usually hand written in C or assem-

W. Müller et al. (eds.),
SystemC: Methodologies and Applications, 247–272.
© 2003 Kluwer Academic Publishers.

bler code, linked with a hand written real time kernel and integrated with the hardware as a final step [K. Shumate, 1992], [Balarin et al., 1997].

Throughout the 1990s the complexity of embedded systems and the time to market pressure increased constantly, thus the previously commented methodology became non-viable. Additionally, implementation technology evolved towards increasingly complex integrated circuits that allowed the integration of the complete embedded system on a chip (SoC) . A SoC normally includes microprocessors, memory blocks, peripheral modules, and hardware components that implement specific system functions (e.g., time-critical processes). A key aspect of the SoC design is the close relationships between hardware and software parts which mean that they need to be designed at the same time (co-design). In order to design these complex embedded systems on time new co-design techniques have been proposed. Originally these were based on a 'describe and synthesize' methodology [Gajski et al., 1994b]. Figure 9.1 shows a classical co-design flow [Balarin et al., 1997] in which from a high level language specification (HardwareC [Gupta, 1995], SpecCharts [Gajski et al., 1994b], Esterel [Balarin et al., 1997],...) the source code (C) of the embedded software is synthesized . Examples of classical co-design tools are Vulcan [Gupta, 1995], Polis [Balarin et al., 1997], Cosyma [Straunstrup and Wolf, 1997] or Castle [Lopez et al., 1993]. In these systems the input description is compiled into models that describe the system behavior. Typically two models are used to describe software: state oriented and activity oriented models [Lopez et al., 1993].

State oriented models use a set of states, transitions between them, and actions associated with states and transitions to model the system. Examples of this category are the CFSM [Balarin et al., 1997], Petri nets [Sgroi et al., 1999] and HCFSM [Gajski et al., 1994b] representations. In order to synthesize the software some representations (strongly oriented to hardware) have to be refined. For example, the CFSM has to be implemented onto an S graph (software graph) before software generation [Suzuki and Sangiovanni-Vincentelli, 1996]. Other representations (e.g., Petri nets, or PN) minimize this problem. The PN model [Sgroi et al., 1999] represents data computation using one type of node (transitions) and non-FIFO channels between computations using another type of nodes (places). Data dependent control (if then else or while do loops) is represented by places, called choices, with multiple output transitions, one for each possible resolutions of the control. Data are represented by tokens passed by transitions through places.

In the activity oriented models the system is described as a set of operations related by data or execution dependencies. They are frequently used in DSP applications. Examples of this model are dataflow graphs and control flow graphs [Gajski et al., 1994b].

After the input description is compiled into one of the above models the hardware/software partition is performed taking into account the design constraints and estimated system performance . After this step the software synthesis begins with the scheduling of the software concurrent tasks on a shared resource: the CPU [Sgroi et al., 1999]. This schedule can be static (defined during the software synthesis process) or dynamic (defined at execution time by a RTOS). The term 'software synthesis' is often used for static schedulers that have to schedule the specification increasing execution speed and reducing the memory usage [Sgroi et al., 1999] [Jiang and Brayton, 2002]. The main disadvantage of these systems is the lack of a code reuse methodology and the problems caused by the input description modification (software debugging difficulties, poor code, etc...).

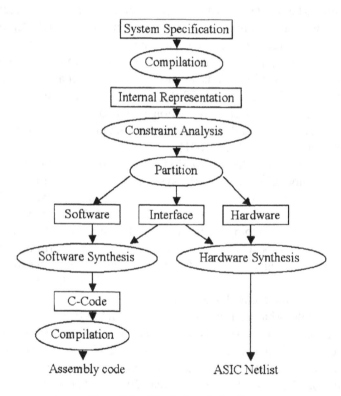

Figure 9.1. Classical co-design flow

In order to generate the embedded processor program the source code (generated after scheduling) has to be compiled with a retargetable compiler [Marwedel and Goossens, 1995]. Today some of the retargetable code generation effort is focused on the porting of open source compilers such as GNU binutils [Abbaspour and Zhu, 2002], or gcc.

Additionally some techniques modify the processor architecture and compiler algorithms in order to improve memory usage, speed, and power performances. After compilation the embedded software has to be downloaded to the system. The code is normally implemented into an expensive 'on chip' memory, thus the code size might seriously affect the cost of the system. To reduce this cost several memory size reduction (e.g., [Kjeldsberd et al., 2001], [Passerone et al., 2001]) and code compression techniques have been proposed, (e.g. [Lekatsas et al., 2000]). The basic idea is to store compressed code into the system memories and decompress the code when it is loaded to the CPU. As embedded processors often include cache memories, specific memory, and speed optimizations for them are required because the correct design of these cache memories has a high impact on system performances. Several techniques have been proposed for improving cache usage [Jain et al., 2001], [Chiou et al., 1999]. These techniques define the cache implementation as well as the necessary software support. Additionally, some techniques can improve memory size with dynamically allocated data structures [Ellervee et al., 2000], increase the memory bandwidth [Gebotys, 2001], and reduce the address calculation [Gupta et al., 2000]. Power consumption is also a very important issue of current systems on chips and some software techniques allow it to be reduced [Fornaciari et al., 2001], [Shiue, 2001].

In the late 1990s the 'describe and synthesize' methodology presented in Figure 9.1 was not able to survive the system on chip revolution, since it did not fulfil the time to market constraints . A new methodology, based on the 'specify–explore–refine' paradigm [Gajski et al., 1994b], was proposed [Chang et al., 1999]. The methodology points out that design reuse and platform based design are the key aspects of the process. A hardware platform is a family of architectures satisfying a set of constraints imposed to allow reuse of hardware and software components [Sangiovanni-Vincentelli, 2002b]. Figure 9.2 shows the design flow of the methodology [Keutzer et al., 2000].

This methodology has been adopted in commercial co-design tools with software generation capability, such as the Cadence Virtual Component Co-Design Environment (VCC) [Cadence Design Systems, 2002] or CoWare [CoWare, 2002]. The basic idea is to assign the functions to be implemented to components of the platform or IP libraries [Villar, 2001].   The system behavior is normally fixed but some methodologies allow implementing different behaviors (adaptive systems) [Ernst et al., 2002]. The mapping is an iterative process in which performance simulation tools evaluate the mapped architecture and guide the next assignment. After mapping, the re-used hardware and software component are added to the design, and the rest of the system is implemented in a similar way to the methodology shown in Figure 9.1.

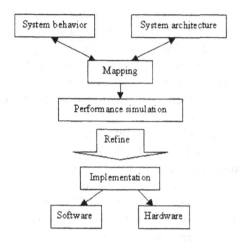

Figure 9.2.   Platform based design methodology

The methodology maps the behavior in a specific platform. It normally in-cludes some pre-designed hardware modules: microprocessor, a fully specified set of buses, memory, and a full specification of a set of I/O devices and inter-rupt controllers. This pre-designed hardware architecture keeps the instruction set architecture (ISA) constant, thus making software porting (or re-targeting) much easier. There are two main concerns for effective platform based design: the software development environment (compiler, linker, assembler, debugger and simulator); and a set of tools that insulate the details of the architecture from the application software [Sangiovanni-Vincentelli, 2002a]. These tools (RTOS and device drivers) make up the Application Program Interface (API Platform) or Platform Program Model [Sangiovanni-Vincentelli, 2002a]. Thus software generation methodologies must take into account this API.

In the methodology shown in Figure 9.2 the role of the system level de-scription language is of great importance and has been highlighted in the 2001 International Technology Roadmap for Semiconductors (ITRS) [Allan et al., 2002]. This report predicts the evolution of the traditional design flow into an integrated framework in which the input is a system level specification and wherein co-design, logical, physical, layout and other tools operate together. In this context SystemC will play a very important role because it allows system level specification (which integrates SW and HW), performance analysis, and verification. Additionally it facilitates the SW analysis and generation, and it is easy to use for engineers with training in C++.

The system model that 'specify–explore–refine' methodologies use consists of a set of processes and communication channels. There is a clear distinction between computation (processes) and communication/synchronization (channels) [Allan et al., 2002]. The process schedule is dynamic and is normally controlled by the platform RTOS. Some commercial [WindRiver, 2002] and academic [Mooney and Blough, 2002], [Vercauteren et al., 1996] RTOS have been developed for embedded systems. These OS have a close relation with the software generation procedure.

A good example is SoCOS [Desmet et al., 2000], a system level design environment which is used for modeling, simulating, system analysis and implementation through gradual refinement. The software generation process begins from the process network of the executable specification. Source code is generated for the processes in which the SoCOS functions which model concurrency and communication/synchronization channel accesses are replaced by functions of a special library: the operating system API (OsAPI). During implementation this OsAPI is replaced by a specific platform RTOS by a simple translation between its calls and the RTOS calls.

A similar approach is used in [Gauthier et al., 2001], [Cesario et al., 2002] in which software wrappers are created during software generation. From an extension of SystemC an architecture analyzer determines the required OS services and a code selector and expander generate a specific RTOS with the required services (software wrapper). During software generation calls to the software wrapper functions implement the RTOS functionality. The problem of the above approaches is that two design representations (original and with 'software wrappers' or OsAPI functions) have to be maintained.

A solution is presented in [Grötker et al., 2002], [Fernández et al., 2002], [Herrera et al., 2003]. The idea is to replace SystemC constructions with RTOS functions. For example, SystemC SC_THREAD processes can be turned into POSIX threads [Grötker et al., 2002]. In [Fernández et al., 2002], the SystemC kernel library is replaced by a new library in which SystemC functions are implemented with RTOS calls. An open source RTOS, eCos [Massa, 2002], is used in [Herrera et al., 2003] to implement the SystemC kernel library. The main advantage of this approach is that the same code can be used in SystemC simulation as well as for platform compilation. It should be pointed out that the support for software modeling in SystemC will be improved in future versions [Grötker, 2002]. Thus it seems that SystemC 3.0 will include new language constructions oriented to software such as scheduler models or dynamic thread and channel creation [Grötker, 2002].

9.2   System Specification Methodology

As was seen in the previous section, the integrated design framework is based on a system level specification from which the design procedures start. The establishment of a specification methodology is key in order to ensure the feasibility and efficiency of all those procedures. It comprises aspects such as the constructing elements, rules, or constraints enabling the inclusion in the specification of features such as funcionality and structure. The model of computation must be well known by the system specifier and taken into account by the implementation and analysis tools.

9.2.1   General Characteristics of SystemC as a Specification Language

As commented previously, complex system specification requires the capability of describing the functionality which can be implemented either in HW or SW. In this environment it is necessary to include some requirements of the language usage [Grötker et al., 2002]. This must cover as many levels of abstraction as possible, ideally all levels of abstraction and computational levels from system to implementation, including HW and SW parts. Additionally an executable specification of the design must be possible, and simulation speed has to be improved over that of traditional HDLs. Finally, another desirable property is the possibility of separating functionality from communication in the design. SystemC accomplishes these requirements when an appropriate specification methodology is used.

In this section a system level specification methodology for the functional part is proposed. This methodology does not affect the test bench for which a standard methodology should be used [OSCI, 2002c].

Structure.   The top level of structure for the executable specification is composed of two main parts: the environment model and the functionality to be implemented (that is, the system). Both parts are instantiated in the *main* function of the global description.

For the system part the main constraint applied is that functionality and communication have to be clearly separated. This important paradigm ensures an optimum implementation. Functionality is encapsulated in functions and processes included in modules, and communication is encapsulated in channels. Modules and processes can only be connected and communicated with channels, even if processes are included in the same module. This restriction is imposed even for communication between the environment and the system.

Figure 9.3 shows an example of the proposed topology. A module can contain other modules and several processes. With this scheme functional and structural hierarchy is supported.

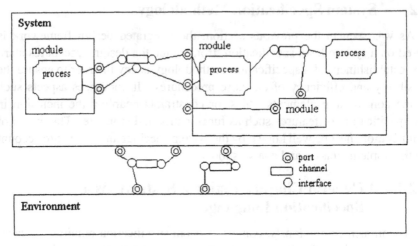

Figure 9.3. System and environment structure

Functionality. System functionality is described in processes which are executed concurrently. In order to ensure the intended separation between communication and functionality no event objects are supported inside processes. The event object is the core object for synchronization and communication in SystemC. With exclusion of these objects, communication takes place out of the process (where only functionality is included). As will be seen later, event objects, will be included in channels which will be in charge of supporting communication and synchronization between processes.

Only one kind of SystemC process is used, the SC_THREAD, which is the most general kind of process. Because no event object is allowed, the notify or wait primitives are not permitted inside processes (and there are no sensitivity lists). There is only one exception: the use of wait(absolute_time), to implement certain timing constraints.

With the above features a process can be represented as in Figure 9.4. There are four kind of nodes: start; wait; channel; and finish. The process execution begins with the start node which represents the sc_start() primitive of SystemC and which is executed in the *main* function. Arrows represent sequential paths of execution (without blocking). These sequential paths can finish in a wait node (wait(absolute_time) primitive), in a finish node, or in a channel method access.

When the process accesses a channel method the simulation context changes to the channel which the process is accessing. This channel can define blocking points or not. The context will return to the process when the channel method returns.

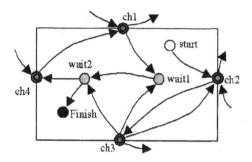

Figure 9.4.   Process representation

A finish node represents the end of process execution, but not the end of the simulation. It is reached when the process closing } or a return primitive inside the process is reached.

Communication.   Communication between processes is performed through channels. The processes can access methods defined in channels and declared in the interfaces of the channel. SystemC, as a C++ language, permits access to object data and methods in different ways, but in order to clearly separate functionality and communication the only way a process can communicate with another process is by accessing a channel to which the other process also has access. When the processes are in different modules the access to the channel has to be performed through ports. This is represented in Figure 9.5 and the following SystemC code.

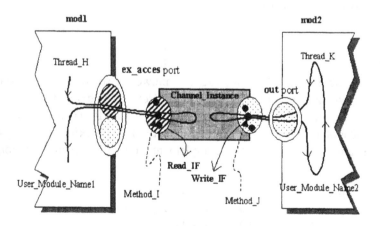

Figure 9.5.   Two process communication through a channel

```
template <class T>
class Read_Write_IF  : virtual public Read_IF<T>,
                       virtual public Write_IF<T> {};

SC_MODULE(User_Module_Name1) { ...
   sc_port<Read_Write_IF>    ex_access; ...
   void Thread_H(void) { ...
      // Method_I belongs to Read_IF interface
      value = ex_access->Method_I(args); ...
   } ...
   SC_CTOR(User_Module_Name1) {
     SC_THREAD(Thread_H); ...
   }
};
SC_MODULE(User_Module_Name2) {...
   sc_port<Write_IF>    out; ...
   void Thread_K(void) { ...
      // Method_J belongs to Write_IF interface
      out->Method_J(args); ...
   }
   SC_CTOR(User_Module_Name2) {
     SC_THREAD(Thread_J); ...
   }
};
   // Channel instances
Channel_Class  Channel_Instance(constructor_args);
   // Module instances...
User_Module_Name1 mod1("mod1");
User_Module_Name2 mod2("mod2");
   // Module port binding...
mod1.in(Channel_Instance);
mod1.out(Channel_Instance);
...
```

In this SW generation methodology a channel library has to be developed. Standard primitive SystemC channels , such as sc_signal, sc_fifo, sc_mutex, sc_mq and RG_protocol, are allowed if software generation oriented implementations are defined for them. Next, as an example, the code for a library channel is presented. This channel defines a FIFO behavior with only one element of any type T. Read and write method are blocking. The way these channels are modified for software implementation will be seen in section 9.3.

```
template<class T>
class uc_simple_channel_SYS : public sc_channel,
           public uc_simple_read _if<T>,
           public uc_simple_write_if<T> {
public:
   void write(const T& data) {
     if (channel_released==false) wait(read_event);
```

```
    data_container = data;
    channel_released = false;
    write_event.notify();
  }
  void read(T& data) {
  if (channel_released) wait(write_event);
    data = data_container;
    channel_released = true;
    read_event.notify();
  }
private:
  T data_container;
  bool channel_released;
  sc_event write_event, read_event;
};
```

9.2.2　SystemC Fundamental Computational Model

Any specification language or notation is based on a certain computational model. Any system can be conceived as a set of interacting components. A computational model is understood as the way in which the system components behave and interact among themselves. The definition of the system components, the set of constraints on them, and the computational model of their interaction constitute a framework [Lee and Sangiovanni-Vincentelli, 1998], [Lee, 2000]. In this Section the fundamental SystemC computational model will be briefly introduced in order to facilitate the understanding of the transformations implied by the proposed SW generation methodology. Terminology from [Lee, 2000] will be used.

The temporal behavior of the processes constitutes essential information throughout the design process. The tag system is one of the most important characteristics of any model of computation. Depending on the tag system used, two main groups of computational models can be distinguished.

Timed Models of Computation.　The general characteristic of timed models is that T is a totally ordered set, that is, for any two events e_1 and e_2 either one precedes the other or they are synchronous (both take place at the same time, that is, $T(e_1) = T(e_2)$. There are several types of timed models. In continuous time systems $T = \Re$, the real numbers. In discrete event systems $T = \aleph$, the natural numbers. In some of these systems transactions in δ time are allowed. In this case the tag set is actually $T = \aleph x \aleph$. Although they are consecutive events $e_1 = (t_1, t_2, v_1)$ and $e_2 = (t_1, t_3, v_2)$, that is, $t_3 > t_2$, they are assigned the same time stamp t_1.

We will call those timed systems in which T represents physical time (seconds) strict-timed systems. We will call those timed systems in which T does not represents physical time (thus providing more implementation freedom) relaxed timed systems [Villar, 2002]. The majority of the languages currently used in HW description, simulation, and design fall into this category. So Simulink, VHDL-AMS and SPICE represent examples of strict, continuous time frameworks. VHDL and Verilog represent examples of strict, discrete event languages. They represent the most typical paradigm in HW design efficiently supporting the design steps from the RT level description down to the gate level implementation. Synchronous languages like Lustre, Esterel, and Signal are examples of relaxed, discrete time languages.

Based on the definition of relaxed, discrete event systems, sequential programming languages like C/C++ or Pascal would fall into this category.

Untimed Models of Computation. The general characteristic of untimed models is that T is a partially ordered set, that is, there are events e_1 and e_2 in which none precedes the other. There are several types of untimed models.

Untimed models of computation are the most appropriate for system specification. They are able to reflect the concurrent nature of the embedded system. Potential problems derived from concurrency, such as non-determinism, can be detected more easily and fixed. An untimed specification is less prone to over-specification, thus providing the designer with a wider design exploration space. In contrast, timed frameworks, specifically, strict-timed frameworks are closer to the final implementation, thus being the most appropriate for system description and analysis at the lower levels of abstraction. Concurrent programming languages like Ada and Java are examples of untimed models of computation.

In asynchronous message passing systems processes communicate by means of dedicated FIFOs. Dataflow models, widely used in DSP design, are based on this computational model. In synchronous message passing systems, like Communicating Sequential Processes (CSP), processes communicate through rendezvous. Lotos and Occam can be included in this framework.

SystemC. The first versions of SystemC were based on a relaxed timed, discrete time computational model including δ time transactions. It was discrete time because there was an underlying master clock from which the other clocks triggering sequential processes and, therefore, advancing time were derived. The main time tag $T_1 \subseteq \aleph$ did not represent physical time. Combinational processes may provoke δ time transactions.

This computational model changed with version 2.0. In order to facilitate the integration of IP blocks modeled at gate level with logic delays, the main time tag represents exact time in sc_time_units. In an opposite direction the synchronization mechanisms wait and notify based on events were added, increasing the flexibility of the communication and synchronization among processes. In its current version SystemC is based on a strict-timed, discrete event computational model including δ time transactions.

Although very similar to VHDL in its underlying computational model, SystemC can be used to model a wider variety of computational models [Grötker et al., 2002]. This is possible for two main reasons. Firstly, the discrete event computational model is simple enough to facilitate its extension to other computational models. This fact had already been achieved through the multiple applications proposed for VHDL outside its original purpose of logic, event-driven simulation. Secondly, the OO features of C++ facilitate the creation of semantic layers above the simulation engine, which can be hidden from the user while implementing a certain model of computation.

The main disadvantage of any strict-timed model is the timing constraint it imposes. Timing flexibility will be possible by allowing the designer to implement computation segments executed in a δ cycle in as much time as required. The need for this flexibility is evident when the corresponding segment has been allocated to SW but it is equally necessary when the segment is allocated to HW. As a consequence the results of the simulation before and after co-design will be different. This is unavoidable in any design step [Villar and Sánchez, 1993]. In order to ensure that the result of the design process corresponds to the previous behavior, a clear methodology, fully understood by the designer, is needed.

9.3   SW Generation

Bearing in mind the increasing design costs, where SW is becoming more important, several methodologies (presented in section 9.1) systematically or automatically generated the binary code to execute over the final platform. In the context of an integrated design framework, a SW generation methodology, starting from the system level and preserving the efficiency of the implementation must be provided.

9.3.1   Single-Source SW Generation

As shown in Figure 9.1, most of the co-design methodologies propose an intermediate step (manual or automatic) that generates the SW source code from the specification code. After that a compilation step produces the binary code downloadable to the platform.

In a single-source approach for the systematic generation of the binary code, the SW synthesis step is eliminated and the same specification code is used for binary code generation as for other design procedures (system level executable generation, co-simulation, performance analysis, etc...) The main advantage of this approach is that it facilitates the SW generation process and debugging, overcoming the need to maintain the coherence between the specification and other intermediate representations.

This approach comes from the observation that the system level specification code and the SW code maintain a high correlation [Fernández et al., 2002]. Moreover, in C-based methodologies system level descriptions and the generated SW code present similar constructing elements in order to support concurrency, synchronization, and communication. For instance, a SystemC process declaration has a direct relation to the corresponding thread declaration in an RTOS API.

For synchronization and interface implementation this relationship is not so direct. SystemC channels normally use notify and wait constructions to synchronize the data transfer and process execution while the RTOS normally supports several different mechanisms for these tasks (interruptions, mutexes, flags, ...). Thus every SystemC channel can be implemented with different RTOS functions. However, this correspondence flexibility (one system channel— several implementations) can serve to give a certain flexibility for an efficient implementation of interfaces, which might strongly affect the performance of the system implementation.

In [Grötker et al., 2002] a solution which requires the substitution of each C++ SystemC construct with an equivalent C code is proposed. Below is an example of the way this substitution is performed:

```
#ifdef SYSTEMC
#include "systemc.h"
#else
#define SC_MODULE(name) struct name
#endif

SC_MODULE(my_module) {
#ifdef SYSTEMC
public: ...
    sc_fifo_out<int>  out; // SystemC port ...
#else
    struct fifo *out;   /* port equivalent in C */ ...
#endif
};
```

In this example the SYSTEMC variable declaration generates the SystemC description, otherwise the C source code is generated by the C preprocessor. The main advantage of this approach is that the use of a C compiler ensures a wide availability of crosscompilers to different microprocessors and an efficient binary code. A high code portability will be obtained if POSIX functions are used to model the RTOS interface.

In [Fernández et al., 2002] [Herrera et al., 2003], a different approach is presented in which the SystemC library elements are replaced by behaviorally equivalent elements of a SW implementation library (SC2RTOS library). This library uses a platform RTOS API and defines some C++ supporting structures. It is the responsibility of the platform designer to ensure the required equivalence between the SystemC functions and their implementation. Since the support library is written in C++ this approach assumes the availability of a C++ cross-compiler. This could reduce portability and efficiency, but it simplifies significantly the specification process. Then, using this technique, the previous code will be as follows:

```
#ifdef SYSTEMC
#include "systemc.h"
#else
#include "SC2RTOS.h" // this is the only code modification
#endif

SC_MODULE(my_module) { ...
  sc_fifo_out<int>  out; ...
};
```

This approach uses a single C++ description for specification and SW code compilation (the main goal of the single source approach), and it will be explained in more detail in the rest of this chapter. The methodology described above has to be extended in order to cover the SW/HW partition. Thus the SYSTEMC variable has to be replaced by two variables: LEVEL which define the description abstraction level and IMPLEMENTATION which specifies the partition (SW or HW) in which the modules are assigned.

The LEVEL variable can take three values: SPECIFICATION, ALGORITHM and GENERATION. When the first value is taken the input code is seen as the system level specification before partition, thus it is a standard SystemC description in which the standard systemc.h library maps the language constructions into the SystemC simulation kernel.

After partition the abstraction level is modified and the LEVEL variable takes the ALGORITHM or GENERATION values. In this case the IMPLEMENTATION variable identifies the modules that have been assigned to SW or to HW.

When the LEVEL variable takes the value ALGORITHM the SW part is executed with the platform RTOS support while the HW part is executed with the SystemC simulation kernel support (thus the HW part is still a standard SystemC description not affected by the LEVEL value). In Figure 9.6 this is represented as two paths, the top path, which represents the SW partition behavior, and the bottom path which represents the HW partition behavior. The full specification will run over the simulation host, although the RTOS can be implemented in different ways. If the RTOS is targeted on the host, a specific implementation

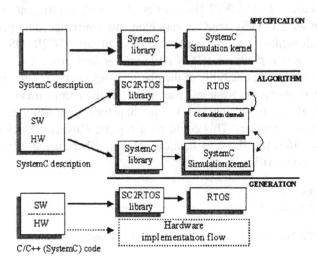

Figure 9.6. Proposed design flow

of the SW/HW and HW/SW communication channels (communicating SW processes and SystemC simulation kernel) has to be developed in order to allow co-simulation. This implementation might be performed by the SystemC simulation kernel itself if a future version of the SystemC supports RTOS model libraries. In this case all the co-simulation will run over the SystemC kernel. If the RTOS is targeted on the platform an instruction set simulator (ISS) will execute the RTOS and SW partition code. In this case driver implementations communicate the ISS and the SystemC simulation kernel.

Finally, the SW partition has to be isolated from the HW in order to enable platform compilation. This is done with the GENERATION value of the LEVEL variable. With this value only those specification constructions of the SW partition (modules of the SW partition and sc_main function) will be compiled for the target platform

9.3.2 Software Implementation Library: SC2RTOS

During software generation the 'meaning' of the code which supports the SystemC constructions in the SW part has to be modified. This is performed at library level and is therefore hidden from the designer. Thus the SystemC user code (after generation, pure C/C++ code) is not modified during the software generation flow. For example, the sc_start() function in the SystemC library will start the SystemC processes, but in the implementation library (SC2RTOS), it will resume all the system tasks under the RTOS control.

One of the main advantages of this methodology is that the development of the
SC2RTOS library is relatively easy because the number of SystemC elements
that have to be redefined is very small. In addition, its file structure can be
similar to the SystemC library structure, thus the porting of the library can be
systematically performed. Figure 9.7 shows the redefined SystemC elements.
They are classified in terms of the element functionality. The figure also shows
the type of RTOS functions that replace the SystemC elements. The specific
RTOS function that replaces the SystemC constructions depends on the selected
RTOS. If a POSIX type API is chosen (for example, EL/IX, [Garnett, 2000])
different RTOS will be supported with the same SC2RTOS library.

		Hierarchy	Concurrency	Communication	
SystemC Elements		SC_MODULE SC_CTOR sc_module sc_module_name	SC_THREAD SC_HAS_PROCESS sc_start	wait (sc_time) sc_time sc_time_unit	sc_interface sc_channel sc_port
RTOS Functions			Thread management	Synchronization management Timer management	Interruption management Memory access

Figure 9.7. SystemC elements to translate

Most of the SystemC constructions can be compiled in a C++ cross-compiler
without an additional library. For example, SystemC C constructions such as
macros, structs, data types, and function declarations, as well as SystemC C++
constructions (data type classes, sc_module class, sc_port template, etc, ...)
can be compiled almost without modification. Additionally, SystemC not only
requires C++ constructions but it also requires object oriented techniques, such
as inheritance, operators overload, etc... For example, in order to declare a
channel class it is necessary to inherit a sc_channel class and an interface
class.

Hierarchy Support. In order to cover hierarchy, the SystemC library provides
some macros (e.g. SC_MODULE and SC_CTOR) and classes (sc_module and
sc_port, sc_module_name, etc, ...). The example below shows a short section
of the SystemC library in which some of these elements are declared.

```
#define SC_MODULE(user_module_name) struct user_module_name : sc_module
#define SC_CTOR(user_module_name) \
    typedef user_module_name SC_CURRENT_USER_MODULE; \
    user_module_name( sc_module_name )
```

```
class sc_module : public sc_object {
    friend class sc_module_name;...
public:
    sc_simcontext* sc_get_curr_simcontext() { return simcontext(); }
    // ETC... The rest not shown
};
```

The implementation library will also declare the same elements, but with a SW oriented implementation. The following example shows the elements of the previous example but declared in the SC2RTOS library. In this case RTOS functions are not needed, so this portion of the library is totally independent of the RTOS.

```
#define sc_module_name uc_module_name
#define sc_module uc_module
#define SC_MODULE(user_module_name) UC_MODULE(user_module_name)
#define SC_CTOR(user_module_name) UC_CTOR(user_module_name)
#define sc_port  uc_port
#define UC_MODULE(user_module_name)\
        struct user_module_name :public uc_module
#if LEVEL == GENERATION
#if IMPLEMENTATION == SW
#define UC_CTOR(user_module_name) \
        typedef user_module_name UC_CURRENT_USER_MODULE; \
        user_module_name(uc_module_name)
#else
#define UC_CTOR(user_module_name) \
        typedef user_module_name UC_CURRENT_USER_MODULE; \
        user_module_name(uc_module_name){}; \
        dummy()
#endif
#endif
class exec_context;
class uc_module { friend exec_context; };
template<class IF> class uc_port{
public:
  void operator()(IF& channel_interface_port); //same hierarchy binding
  void operator()(uc_port<IF> &parent_port); //child-parent port binding
  IF *access(); IF *operator->(); ...
private:   // ... Class implementation not shown
};
```

This example partially shows an implementation for a module granularity level. For SW generation the constructor of HW modules is an empty function, thus their processes are not declared. With the correct compiler optimization options, HW process functions are not included in the compiled code. This technique can be translated to the process SC_THREAD macro (process granularity level). In this case, each process declaration is eliminated and the user is not obliged to use the SC_CTOR macro for the constructor declaration.

```
#define SC_THREAD(func) UC_THREAD(func)
#if LEVEL == GENERATION
#if IMPLEMENTATION == SW
#define UC_THREAD(func) \
        declare_thread_process(UC_CURRENT_USER_MODULE,func)
#else
#define UC_THREAD(func)
#endif
#endif
```

Concurrency Support and Execution Control.   Some SystemC elements are dedicated to supporting concurrency. In this methodology, the SC_THREAD macro is used for this purpose. The implementation library will use RTOS API calls for the concurrency support, so the SystemC threads have to be mapped to the underlying RTOS threads. In addition to the SC_THREAD macro some kinds of structures are needed to maintain the process list, where all data necessary for the management of each process (handlers, process stack pointers, etc...) are stored.

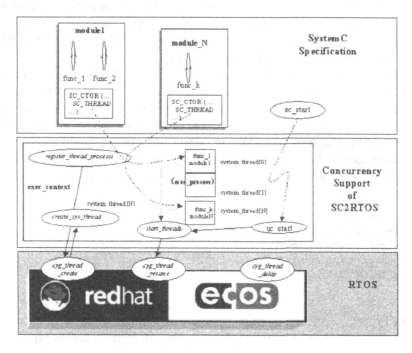

Figure 9.8.   Concurrency support of the implementation library

For example, Figure 9.8 shows the eCos implementation of the SC2RTOS (SC2eCos library). This library has an object (execution_context) functionally equivalent to the sc_simcontext class in the SystemC library. This object maintains a list in which every element stores the data that an eCos thread requires to execute. The execution context provides a method (register_-thread_process) which registers a new system process in the process list, allocating the necessary data structures and calling the eCos process declaration call (cyg_thread_create). The method is called during SystemC process declaration by means of the redefined SC_THREAD macro. Thus, this implementation is transparent to the user who is still using SystemC syntax. The execution_context object has another entry function (uc_start) which starts the execution of all system threads by using the eCos cyg_thread_resume call. The sc_start macro is also redefined in order to call uc_start with a SystemC syntax.

Timed Specification.   In SystemC it is possible to include some time information at specification time. The process execution can be halted during a specific time period by using the wait(sc_time) SystemC primitive. Once that period expires the process resumes without needing any external event.

A SW implementation of this feature can be given by directly programming one of the timers of the target system. Nevertheless, in order to widen the SC2RTOS library targetability it is normally better to make use of the delay services offered by the underlying RTOS API.

Going into detail, the implementation library will include the sc_time_unit type, the sc_time class and the wait(sc_time) function. The wait(sc_time) function will call the RTOS function for the thread delay service (for example, sleep() in POSIX or cyg_thread_delay in eCos), providing it with the correct delay parameter (normally, in system ticks, or equivalent units) calculated from the sc_time parameter and the system clock period. The following example shows the wait implementation in the SC2eCos library:

```
enum sc_time_unit {SC_FS = 0, SC_PS, SC_NS, SC_US, SC_MS, SC_SEC};
#define wait uc_wait
inline void uc_wait( double value, sc_time_unit tu) {
  unsigned int tick_count;
  // calculation of tick_count...
  cyg_thread_delay(tick_count);
};
```

Data Types.   Another question is the software implementation of the object types. The specification may include integer and floating point C++ types (e.g., int, float, double) as well as SystemC fixed precision integer (e.g., sc_int, sc_uint) and fixed point types (e.g., sc_fixed). The C++ types are directly compiled but the SystemC types have to be defined. One possibility is to use the SystemC data type implementation. This solution maintains intact the data behavior of the system level specification. Another possibility is to replace the

SystemC implementation with a SW oriented implementation that replaces the underlying SystemC implementation (e.g., 64 bit integer for fixed precision) with a simpler implementation. In order to obtain an efficient SW code, the best option is to use C++ types at specification level.

9.4 SW/SW Communication and Driver Generation

As outlined in the specification section, channels are the basic components used to connect processes and modules. Any other communication primitives can always be seen as a particular case of a channel. The system specification uses SystemC primitive channels; thus it is necessary to provide a library that includes the SW oriented implementation of these channels. This library may include the standard SystemC distribution and platform specific channels. If the designer defines his own channels he should provide the SW oriented implementations.

Every primitive channel can have three types of SW oriented implementation: SW/SW; HW/SW; and SW/HW channel. A SW/SW channel communicates processes of the SW partition. Thus specific RTOS services for process communication are used to implement it. The HW/SW and SW/HW channels communicate processes found in different partitions. The SW partition will include some RTOS functions (drivers) that communicate with the HW part. This part has to include some logic which implements the HW part of the channel. The type of implementation could be specified by the designer or automatically generated by a source code analyzer. Additionally every type can have different codings depending on the RTOS communication services.

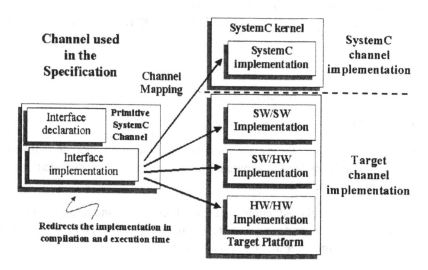

Figure 9.9. SystemC channel implementation

Figure 9.9 shows a possible channel library implementation. The main idea consists in describing the channel class as a wrapper from where the implementation class is called (this process can be called 'channel mapping'). A function pointer C technique can be used for channel mapping. Another possibility is a C++ technique consisting in an interface class object pointing to the implementation class. The implementation is selected depending on a variable (with values SW_SW, SW_HW, or HW_SW) which can be provided manually (as a parameter of the channel class constructor method) or detected automatically. Each channel implementation class uses communication mechanisms provided by the underlying RTOS. For example, in the SC2eCos library the synchronization can be implemented through several RTOS objects (flags, mutexes, mailboxes, etc...). A highly suitable mechanism is the eCos flag, which enables a quick development of the implementation library, since it is a primitive similar to SystemC events. While in general the SW/SW channel implementation only depends on the RTOS API, the HW/SW or SW/HW channel implementation is also dependent on the platform architecture. This is because the methods of the HW/SW and SW/HW implementation classes will access the specific and generic peripheral HW of the platform. In [Fernández et al., 2002] an approach is given using memory mapping and interruption techniques to implement channel synchronization and communication mechanisms.

9.5   Example of Software Generation in SystemC 2.0

Figure 9.10 shows a single source SystemC code example. The top hierarchical module (described in file top.cc) and one of the hierarchical modules (described in file ModuleN.h) are presented.

Some assumptions affect the way the LEVEL and IMPLEMENTATION variables are used:

- The platform has only one processor;

- The partition is performed at module level in the top hierarchical module;

- Every module is assigned to only one partition;

- Hierarchical modules are assigned to the same parent module partition. These can be defined in separate files, but they must be included in the parent file during compilation;

- Only the channels instantiated in the top hierarchical module can communicate processes assigned to different partitions. This is a consequence of the assignment of the hierarchical modules.

These restrictions allow the HW/SW partition to be specified by means of #define statements placed before the modules they assign (e.g. module1 is

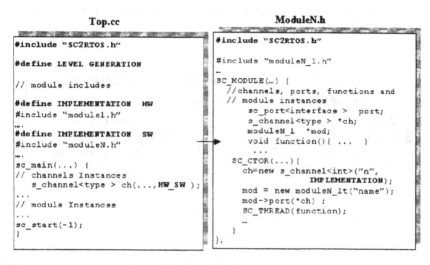

Figure 9.10.   SystemC description example

assigned to the HW part). All the statements which introduce additional information in this description (as partition) have been highlighted with bold fonts. For example, the `general.h` file will include (according to the LEVEL parameter), the SystemC library (`systemc.h`) or the support library SC2RTOS (`SC2RTOS.h`). The LEVEL variable is defined in the `top.cc` file, but it could also be introduced as a compilation switch or a variable in the `Makefile` compilation script. As shown in Figure 9.10, the implementation library has to be included before a module is declared. This allows the compiler to use the correct implementation of the SystemC constructions in every module declaration. Every specific instance of a channel defines an additional argument that specifies the type of communication (HW_SW, SW_HW, SW_SW, or HW_HW).

A simple design example was developed in [Herrera et al., 2003] in order to evaluate these techniques. This example consisted in a car Anti-Lock Braking System (ABS) [CoWare, 2002] example, with about 200 SystemC code lines and including 5 modules with 6 concurrent processes. The system was implemented in an ARM based platform that included an ARM7TDMI processor, 1Mbyte RAM, 1Mbyte Flash, two 200 Kgate FPGAs, a small configuration CPLD and an AMBA bus. The open source eCos operating system was selected as embedded RTOS.

In order to generate software for this platform–OS pair, the SystemC to eCos (SC2eCos) Library was defined. This library is quite small, about 800 C++ code source lines, and it basically includes the concurrency and communication support classes which replace the SystemC kernel (see Figure 9.11). This library uses the basic techniques previously explained (it also includes a channel library with its corresponding implementation library) and could be easily adapted to a different RTOS API or platform.

	Thread management	Synchronization management	Interruption management
eCos Functions	cyg_thread_create cyg_thread_resume cyg_user_start cyg_thread_delay	cyg_flag_mask_bits cyg_flag_set_bits cyg_flag_wait	cyg_interrupt_create cyg_interrupt_attach cyg_interrupt_acknowledge cyg_interrupt_unmask

Figure 9.11. eCos functions called by the SC2eCos library

In this work the resulting memory layout was also obtained. One of the most interesting conclusions was that there was a minimum memory size of 53.2 Kbyte that could be considered constant (independent of the application). This fixed component could be divided in two parts: a 31 Kbyte contribution owed to the default configuration of the eCos kernel, which includes the scheduler,interruptions, timer, priorities, monitor and debugging capabilities, and a 22.2 Kbyte contribution, necessary for supporting dynamic memory management, as the C++ library makes use of it. The variable component mainly increases with the introduction of the new system functionality.

9.6   Impact of SystemC 3.0 Release

Since its first version SystemC has been evolving towards improving its capabilities of system level specification and design. The earliest releases of SystemC were mainly focused on the support of traditional HDL features, specifically RTL description, simulation, and synthesis. SystemC 2.x made possible orthogonality between communication and functionality by providing process communication through channels and more abstract data types. The OO features of C++ made it possible to hide the underlying strict-timed, discrete event MoC by creating semantic layers able to emulate different, more abstract models of computation. These are the fundamental features allowing the use of SystemC 2.x as a system level specification language supporting HW/SW co-design.

The SW generation methodologies described in this chapter are based on these improved capabilities. Nevertheless, there are still some improvements for SW modeling, which have been highlighted [Grötker, 2002] and which will be incorporated in the next release of the language, SystemC 3.0. Most of these new features are oriented specifically to SW modeling and efficient SW generation. Another important aspect is related to the SystemC scheduler. Up to the present this scheduler has some limitations for the specification flexibility:

- It is non-preemptive. The process executes until a wait() is reached. Only then, the control is passed to the scheduler;

- It cannot take into account resource sharing (where a resource can be a processor or a bus). In SystemC 2.x any process is executed in parallel with others, thus sequential execution can only be obliged by synchronizations. Thus the sequential execution of several processes in a processor cannot be directly modeled;

- It defines no specific order for the choice of a process of the ready to execute queue.

All this means that a SystemC execution does not represent the actual execution on a specific target. This makes performance analysis (for example, the schedulability analysis) difficult whenever a real target is considered by means of its full characterization.

In order to overcome this problem SystemC 3.0 will provide several mechanisms as consume(sc_time) to model the consumption of processor resource (forcing sequential execution in that resource), and a hierarchy of *scheduler models* to enable the consideration of different scheduling policies of the RTOS platform without any change in the thread code. Additionally, new facilities such as dynamic process creation and process control (suspend, resume, kill,...), dynamic creation of primitive channels and a library for typical SW communication (Message queue, Mailbox, etc, ...) will be added.

From the point of view of software generation, in [Grötker, 2002] it is mentioned that "SystemC 3.0 will not provide a generic RTOS API and/or a generic processor model, but the language constructs to implement those". This means that the development of SW generation methodologies from SystemC such as those explained here, are necessary and that SystemC 3.0 will be flexible enough to implement a complete SW generation methodology, overcoming the main lacks pointed out in this section.

As a consequence the number of elements included in the SC2RTOS library will increase with respect to those shown in this chapter. For example, it will be necessary to include the use of the RTOS API functions for dynamic process management in order to implement the SystemC dynamic process creation and control. Attending to the scheduler models, a non-trivial mechanism will be required for the mapping from the scheduler mode plus its set of attributes to the specific RTOS mechanism that allows the scheduling policy to be controlled or defined if possible.

In any case, the SW generation methodology described in this chapter will remain the same showing that SystemC is a powerful language for system level specification and modeling, system profiling, and embedded SW generation on any RTOS.

9.7 Conclusions

This chapter shows the existing approaches for embedded software generation, focusing on a single-source approach from SystemC for platform based design. The methods for SW generation based on SW synthesis have been outlined and the single-source approach as an efficient embedded software generation method based on SystemC has been presented. Focusing on the single-source generation method, the required specification methodology, the MoC, and the Design Framework, where this SW generation technique must take place, have been described. This design strategy allows the designer to concentrate the design effort on a single specification, thus preventing errors, raising the abstraction level, allowing early verification, and therefore gaining in productivity.

SW generation from SystemC can be simple based on the substitution and redefinition of SystemC class library construction elements. Those elements are replaced by typical RTOS functions and C++ supporting structures. The use of a C++ cross-compiler serves to optimally reach the single-source paradigm while a C cross-compiling strategy might achieve the optimum code size and speed efficiency. One important advantage is that the method is independent of the selected RTOS and any of them can be supported by simply writing the corresponding library. This targeting can be made wider if a generic RTOS API such as POSIX is used.

Experimental results for the C++ crosscompiling approach demonstrate the quick targeting and efficient code generation, with a minimum memory footprint of 53.2 Kbyte when the eCos RTOS was used. The real overhead with respect to a pure eCos implementation is relatively low, especially taking into account the advantages obtained: support of SystemC as unaltered input for the methodology processes; reliable support of a robust and application independent RTOS, namely eCos, which includes an extensive support for debugging, monitoring, etc, ...

Finally, it is shown how the current evolution of the language will affect this methodology. In general, it will evolve in coherence with the concepts explained, easing the inclusion of some of them, for example, the development of RTOS emulation libraries over SystemC 3.0 will facilitate cosimulation at algorithmic level using only the SystemC kernel

Chapter 10

SystemC-AMS:
Rationales, State of the Art, and Examples

Karsten Einwich[1], Peter Schwarz[1], Christoph Grimm[2], Christian Meise[2,3]

[1] *Fraunhofer IIS/EAS, Dresden, Germany*

[2] *Professur Technische Informatik, University of Frankfurt, Germany*

[3] *Continental Teves AG oHG, Germany*

Abstract   Many technical systems consist of digital and analog subsystems in which some of the digital parts are controlled by software. In addition, the environment of the systems which have to be designed may comprise analog components. Mixed-signal simulation, i.e. the combined digital and analog simulation, is very important in the design of such heterogeneous systems. Unfortunately, state of the art mixed-signal simulators are orders of magnitude too slow for an efficient system simulation. Furthermore, the co-simulation of mixed-signal hardware with complex software, usually written in C or C++, is insufficiently supported.

A promising approach to overcoming these difficulties is the application of SystemC and its extension to the Analog and Mixed-Signal domain (SystemC-AMS). The available SystemC 2.0 class libraries permit the simulation of digital systems at high levels of abstraction. The generic modeling of communication and synchronization in SystemC allows designers to use different models of computation appropriate for different applications and levels of abstraction. Therefore the SystemC 2.0 class libraries can be extended easily to analog and mixed-signal simulation by the implementation of new classes and methods.

In this chapter the requirements for the analog and mixed-signal extension of SystemC coming from important application areas (telecommunication, automotive, wireless radio frequency communication) are analyzed. For these application domains we present simple but efficient solvers written in SystemC 2.0. Finally, we give a brief overview of the first prototype implementation of SystemC-AMS.

W. Müller et al. (eds.),
SystemC: Methodologies and Applications, 273–297.
© 2003 *Kluwer Academic Publishers.*

10.1 Introduction

Many technical systems consist of digital and analog subsystems in which some of the digital parts are controlled by software. Which modeling languages and simulators can be applied most effectively in the design process of such systems? VHDL, VerilogHDL, and SystemC are the best known candidates for this task but they have specific deficiencies.

SystemC supports a wide range of Models of Computation (MoC) in the digital domain and is very well suited to the design of digital HW/SW systems. Different levels of abstraction (from functional level down to register transfer level) may be considered in the refinement process during the first system design steps. In many applications the digital HW blocks and SW algorithms interact with analog system parts and the continuous time environment. Owing to the complexity of these interactions and the influence of the analog parts with respect to the overall system behavior, it is essential to simulate the analog parts together with the digital ones. Furthermore, an efficient analysis of the analog system parts will become increasingly important with smaller chip structures (deep submicron effects) and reduction of supply voltages. This requires an integrated system design and verification environment (from specification level down to circuit level). Higher levels of abstraction should be preferred to save computation time and lower levels have to be used for other subsystems to achieve accurate simulation results.

VHDL-AMS and Verilog-AMS are modeling languages well suited to design tasks like analog circuit design, design of small blocks like PLLs or A/D converters, and other mixed-signal circuits with high accuracy requirements but with a low complexity. However, for the design of very complex integrated circuits (including several processors, embedded software, other digital hardware, and analog blocks) these languages lack of concepts for the description on higher levels of abstraction. There are no possibilities of increasing the simulation performance significantly by the use of application oriented, optimized models of computation. The coupling with software (written in C or C++) is cumbersome and leads to an additional decrease of simulation performance. Tools like Matlab/Simulink provide the user with powerful possibilities for the description of analog and mixed-signal systems at system level, but they do not support modeling of digital hardware/software systems or the simulation of analog subsystems on the electrical circuit level.

To fill the gap, this chapter introduces application specific methods for the description of analog components embedded in a large system consisting of complex digital hardware and software blocks. The main idea is the application of SystemC and its extension to Analog and Mixed-Signal problems. This leads to the proposal of 'SystemC-AMS'. The SystemC simulation environment provides class libraries which can be used for the simulation on different levels of abstraction. These class libraries can be extended to analog and mixed-signal simulation by the implementation of new classes and methods. In the analog domain various models of computation can be realized by the implementation of different analog solvers.

In the following section we first give a very brief introduction to the modeling and simulation of analog systems. Then we analyze the requirements for the analog and mixed-signal extension of SystemC coming from important application areas (telecommunication, automotive, wireless radio frequency communication). Various classes of signals and systems (e.g., time discrete, time continuous, conservative, non-conservative) have to be considered. The synchronization methods between analog and digital simulation are discussed under the aspects of implementation of analog solvers. The application areas show special features which have to be analyzed carefully to define the most appropriate models of computation and the selection of analog solvers.

In the next sections we give an overview of a first prototype implementation of a mixed-signal simulation environment using the currently available SystemC 2.0 version. Different modeling examples illustrate the SystemC extension to the analog and mixed-signal domain by using new classes and methods for behavioral models as well as netlist descriptions. A layer model for a very flexible extension of the existing SystemC kernel is proposed in the last section. It offers many options for embedding analog solvers: from simple, self-programmed solution algorithms for special classes of linear networks up to highly sophisticated solvers for nonlinear differential equations or coupling with commercially available analog and mixed-signal simulators.

10.2 Modeling and Simulation of Analog Circuits and Systems

Mixed signal systems consist of analog and digital sub-systems (modules, blocks) and signals processed by these modules. In this chapter we use the word 'signal' in a traditional sense as a 'carrier of information', independent of the special meaning of the keyword 'signal' in modeling languages like VHDL-AMS or SystemC.

Digital modules can be modeled and simulated by the communication of discrete processes. These processes communicate with each other and compute output signals only at discrete points in time, and are activated by events. The signals in such systems are assumed to be piecewise constant (see figure 10.1, discrete event signal). Discrete processes can be used to model systems from gate level up to complex software systems. These processes and systems can be described and simulated in an efficient way with SystemC.

Analog systems are specified by differential equations (DE) or, more generally, by differential-algebraic equations (DAE) . They can be classified into conservative and non-conservative systems. Signals in *non-conservative* systems are directed. In an abstract sense they model the directed communication between analog modules. For example, block diagrams represent non-conservative systems.

The signals in *conservative* systems represent physical quantities associated with an across value and a through value. In the case of an electrical network the across value represents the voltage between a wire and a reference node and the through value is the current through a wire. Thus Kirchhoff's laws have to be satisfied. Signals in conservative systems are bidirectional.

Unlike discrete signals, analog signals (in both conservative and non-conservative systems) are not assumed to be piecewise constant. In general they consist of piecewise differentiable segments; the value of an analog signal changes continuously with (continuous) time as shown, e.g., in figure 10.1. For simulation we approximate such signals by, for example, a linear interpolation between sample points.

We furthermore distinguish between *static* and *dynamic* modules. A module is static if it models a purely combinational dependency between inputs and outputs. An example for a static module is the behavioral model of a very idealized operational amplifier with frequency-independent gain and voltage limitation. A module is dynamic if we need differential equations to model its behavior. An example of a dynamic module is a low-pass filter.

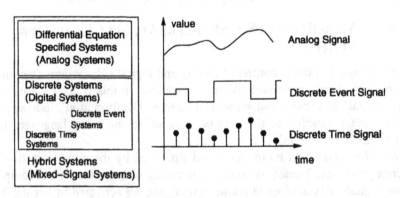

Figure 10.1. Classes of systems and signals

For all these analog systems there are no special constructs in SystemC to model the analog behavior. The distinction of the different types does not cover all situations which arise in modeling and simulation of mixed signal systems.

Analog modules are often designed in such a way that the signals at their interfaces are non-conservative (directed) signals whereas the internal signals of the models are conservative. This design style makes the combining of modules easier to understand because the interactions between the modules are assumed to be unidirectional.

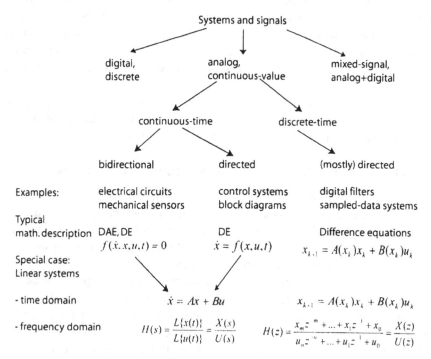

Figure 10.2. Taxonomy of systems and signals

In other situations analog blocks are described by *differential* equations in a first modeling step. In some application areas, such a block has a behavior which is relatively simple to describe and the system of differential equations can be solved by application of explicit or implicit numerical integration formulas with constant time steps. These integration formulas may be implemented as a part of the model. In this second modeling step we obtain a model of the analog block in form of *difference* equations (see section 3.1). If the signals at the input and output terminals of the analog block may be considered to be non-conservative we have a simple behavioral description in the form of a system of difference equations. SystemC has enough possibilities for describing and simulating such analog blocks. Of course, this approximation may be used only for non-stiff differential equations (see section 3.2) and for models without complicated transistor models.

Figure 10.2 summarizes the taxonomy of analog systems and signals discussed. The distinction between these and many other, more complicated, cases that are not discussed here is not an academic one! In real mixed-signal systems we will find all of these block types. It is a big disadvantage of the conventional analog simulators that they are not able to differentiate between these block types and not able to select the most appropriate simulation algorithms for each block. Instead they are oriented on the general case: stiff nonlinear differential-algebraic equations to be solved with very small time steps. This decreases the simulation speed by orders of magnitude compared with a more flexible simulation strategy. It is a main topic of this chapter to show that an extension of SystemC to SystemC-AMS gives the basis for such a more effective simulation strategy.

The implementation of the different model types and simulation algorithms in SystemC is described in the next sections, together with additional considerations of application specific problems (e.g., telecommunication, automotive).

10.2.1 Simulation of Analog Modules and Systems

For the simulation of analog modules we have to set up a system of differential-algebraic equations (DAEs). Several methods of setting up these equations are known. An important example is the Modified Nodal Analysis (MNA) which is used to set up equations of electrical networks. A DAE system can be solved by various numerical integration techniques. Usually a dynamic time step control is applied. This time step control reduces or enlarges the time step with respect to an estimated error. Such dynamic time step control algorithms need a backtracking to previous simulation time points. If we have an analog system which consists of a number of closely coupled analog modules, a global set of equations has to be set up and solved globally (see also chapter 11). In this case each module inserts its equations into the global set of equations. This technique is used in most standard simulators, such as Spice, Mixed-Signal- VHDL-AMS and Verilog-AMS simulators. It is reliable and precise, but unfortunately very time consuming.

Non-conservative systems are often solved by ordering the computation sequence in the direction of the signal flow between the modules. The modules then compute their behavior locally (object oriented), instead of inserting their equations into a global system of equations. In this case the equations of each module can be solved independently using an appropriate algorithm. However, in the case of combinational loops (loops without delays) such an algorithm has to assume an initial value and then iterate until the computed value can be considered to be stable. For such an algorithm all modules inside a loop must be able to backtrack to the last time point.

The DAE solvers of standard analog or mixed signal simulators often lead to non-acceptable simulation times for system level simulations of larger mixed-signal systems. Some restrictions for the system of equations and the time step control can speed up the simulation by orders of magnitude. Possible simplifications for non-conservative systems are:

- Splitting into linear dynamic and nonlinear-static modules;

- Using constant time steps;

- Avoiding zero delay loops; (i.e., avoiding iterations).

These simplifications may be applied to various blocks alone or combined together, and are a result of a careful investigation of the different blocks, their behavior, and their interactions. Furthermore, such simplifications simplify the modeling of mixed-signal systems using SystemC 2.0.

10.2.2   Synchronization

Analog and digital modules communicate and exchange data via signals at their ports or terminals. Therefore the digital (discrete event) simulator and the analog solver have to be synchronized. In general two directions of synchronization have to be considered:

- A digital event changes an analog value, or a discrete process samples an analog value;

- An analog value crosses a threshold and creates a digital event.

In the first case a digital event initiates the synchronization at a known point in time. For the second case, the analog simulator has to calculate the exact time of threshold crossing and to schedule at this time a digital event. In VHDL-AMS such mixed-signal simulation cycle is part of the standard (IEEE 1076-1). In the SystemC simulation environment this synchronization is not yet prepared. A non-effective synchronization scheme can slow down the simulation performance significantly, e.g., by repeated backtracking of the analog solver.

At system level the communication between analog modules can be modeled very often with non-conservative signals. Such non-conservative modeling of the interaction between modules is independent from the description of the module itself, which can also be a conservative system.

If a system is oversampled a delay T of one time step in feedback loops can be accepted. In this case we can compute the outputs of an analog system from known inputs in the signal flow direction without iterations. Although this results in a limited precision, it is a pragmatic rule that an oversampling ratio of 10 leads to a similar accuracy as with circuit simulators such as SPICE or VHDL-AMS. This type of synchronization has many advantages for system level simulation:

- We can simulate each analog module independently of each other. This allows us to use problem specific simulation methods in each module;

- In comparison with more complex synchronization schemes no back-tracking is required;

- The synchronization scheme corresponds to the single rate static data-flow Model of Computation. This model of computation can be easily implemented in SystemC 2.0 using the sc_fifo primitive channel of size one.

The sc_fifo uses a blocking read and write interface. A call of the read method removes a sample from the FIFO and returns the value. A call of the write method adds a sample value to the FIFO. The read method suspends the calling process until a value is available. The write method suspends the calling process until a value can be written into the FIFO.

In this way the analog modules will be simulated in signal flow order. Note that if one out port drives more than one in port, this simple data flow realization needs a fork module which copies the in port value to several out ports. This can be simplified with the implementation of a specific SystemC channel.

Each module has to compute its dynamic or static function independently from all other instances only using the output values of the previously computed instances as input, and its internal state. For example, we can define an analog signal class sca_quantity as follows:

```
//defining analog ports and channels
typedef sca_quantity      sc_fifo<double>      //channel
typedef sca_quantity_in   sc_fifo_in<double>  //in-port
typedef sca_quantity_out  sc_fifo_out<double> //out-port
```

This signal class models a directed signal flow in non-conservative systems, similar to the port quantities in VHDL-AMS. Then we can use it in modules as follows to declare directed (non-conservative) inputs or outputs of an analog module:

```
SC_MODULE(my_module)
{
    sca_quantity_in  input;
    sca_quantity_out output;
    ...
}
```

As a very simple example of an analog module the following code shows the implementation of a sinusoidal source:

```
SC_MODULE(sin_src) {
  sca_quantity_out output;
  double Ampl, Freq, T;      //parameter
  void sig_proc() {
    while(true) {
      //sc_time_stamp().to_seconds() returns SystemC time
      output.write(Ampl*sin(2.0*M_PI*Freq*sc_time_stamp().to_seconds()));
      wait(T,SC_SEC);         //the process is suspended for time T
    } }
  SC_CTOR(sin_src) {
    SC_THREAD(sig_proc); //process registration
  }
};
```

This signal source introduces the time, which is, in contrast to the 'untimed' static data-flow model of computation, mandatory for modeling analog signals. A delay of one time step will be realized by writing the out port sample before reading the in port sample. The delay time is determined by the time distance between the samples, which were generated by the signal sources. An explicit delay module can be realized as follows:

```
SC_MODULE(delay) {
  sca_quantity_in  input;
  sca_quantity_out output;
  double init_value;      //parameter
  void sig_proc() {
    double state_value=init_value;
    while(true){
      //first time, the init_value is written:
      output.write(state_value);
      state_value=input.read(); //read value will be written next time
    } }
  SC_CTOR(delay) {
    SC_THREAD(sig_proc); //thread process registration
} };
```

Note that we can also set the value init_value in order to determine initial conditions. The definition of non-conservative analog signals shows that the extension of SystemC to analog and mixed-signal systems does not need new language constructs, as for example in VHDL-AMS compared with VHDL.

Instead we add new methods or classes to the existing SystemC class library. In the following we describe methods that allow us the efficient simulation of analog modules.

10.3 Problem Specific Simulation Techniques

In section 2 we have discussed properties of analog systems from an abstract point of view (conservative and non-conservative systems, directed and bi directional signal flow, modeling by differential and difference equations). This section describes application specific properties of systems to be simulated. Methods can be developed which use these properties and allow us the simulation of such systems using SystemC 2.0 with some simple extensions (see also [Bonnerud et al., 2001]). We consider the following types of systems:

- Signal processing and over sampled systems — In these systems the time step is constant. Methods known from the design of digital filters can then be applied to create discrete time models of the analog (continuous time) circuits;

- Reactive and stiff systems — In stiff systems very different time constants can be found. Therefore an analog module can be inactive or have only very slow changes of states and signals for a long period in time. However, under other conditions the signal values will change rapidly. Therefore numerical integration and synchronization with constant time steps is not appropriate. In reactive systems discrete processes are typically invoked by conditions which occur at arbitrary points in time;

- RF systems — In radio frequency systems a simulation in the time domain leads to very high simulation times. Therefore base band modeling techniques are used.

Note that many systems combine properties of these classes.

10.3.1 Signal Processing and Over Sampled Systems

Typical examples for signal processing systems are telecom applications [Einwich et al., 2001], [Zojer et al., 2000]. Most telecom applications consist of analog or digital filters, some linear network elements, and digital control algorithms. Usually the design of such systems starts with the specification of simple block diagrams. For the analog/digital conversion sigma–delta converters are usually used. Furthermore, such systems are mostly oversampled. Which means that the time constants are much lower than the sampling period. Therefore a constant time step integration and synchronization is sufficient.

For system level investigations we can split analog modules into a linear dynamic and a nonlinear static part. The linear dynamic part can be solved by a very simple and fast linear integration algorithm with constant time steps. Static nonlinear functions can be modeled directly with SystemC. It makes no difference if they are computed in a discrete event or in a continuous time model of computation.

Numerical integration. For the simulation of linear dynamic parts integration algorithms such as Euler's formula or the trapezoidal rule are applicable. The following equation demonstrates the application of the backward Euler integration algorithm to linear DAEs in the form:

$$A\dot{x} + Bx + q(t) = 0, \tag{10.1}$$

A and B are square matrices, x is the state vector, and q is a vector of independent variables. Equation 10.1 can also be used for the computation of transfer functions, pole zero or state space representations. With the backward Euler formula and the time step $T = t_n - t_{n-1}$:

$$\dot{x}(t_n) = \frac{x(t_n) - x(t_{n-1})}{T}$$

we obtain the following system of equations:

$$x_n = Z^{-1}(-q(t_n) + Wx_{n-1})$$
$$\text{with}$$
$$Z = \frac{1}{T}A + B \quad \text{and} \quad W = \frac{1}{T}A.$$

If a constant time step T is used Z^{-1} and W have to be computed only once in the initialization phase. If Z is not a regular matrix a more complicated algorithm must be used.

Filter transformation and identification. A linear system whose output signals are computed with constant time steps, can be considered as a digital filter. Therefore filter design techniques like bilinear transformation and filter identification can be used also. For the filter design approach Matlab, for example, provides functions such as **bilinear** or **invfreqz**. These functions compute the coefficients for a discrete time transfer function. The resulting digital filter can be implemented easily in SystemC. Compared with the numerical integration approach the filter identification optimizes the frequency domain behavior of the digital filter to fit the behavior of the analog transfer function as much as possible.

The following example demonstrates the modeling of a low pass filter, realized with an operational amplifier (figure 10.3, left).

We obtain a discrete time behavioral model of the analog low pass filter as follows:

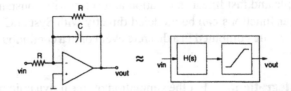

Figure 10.3. Low pass filter and its simple non-conservative behavioral model

- First, we must transform the conservative model of the low pass to a non-conservative model. Furthermore, we split the filter model into a linear dynamic and a nonlinear static part. Figure 10.3 shows the resulting non-conservative model;

- In a second step we set up a continuous time transfer function $H(s)$:

$$H(s) = \frac{v_{out}(s)}{v_{in}(s)} = \frac{-1}{1 + RCs}.$$

- Then we apply the bilinear transformation:

$$s = \frac{2}{T} \frac{z-1}{z+1}$$

to obtain a discrete time transfer function $H(z)$ which approximates the behavior of the analog transfer function $H(s)$:

$$H(z) = \frac{-1}{1 + \frac{2RC}{T} \frac{z-1}{z+1}} = \frac{-(1 + z^{-1})}{\left(1 + \frac{2RC}{T}\right) + \left(1 - \frac{2RC}{T}\right) z^{-1}}. \qquad (10.2)$$

Note that T is the constant time step and z^{-1} is a delay of one time step.

- Equation 10.2 is a discrete time transfer function $H(z)$ with the following filter coefficients:

$$B_0 = -1 \qquad\qquad B_1 = -1$$
$$A_0 = 1 + \frac{2RC}{T} \qquad\qquad A_1 = 1 - \frac{2RC}{T}$$

In the time domain we can compute the output voltage v_{out} from the input voltage v_{in} by the following difference equation:

$$v_{out} = \frac{1}{A_0}\left(B_0 v_{in} + B_1 z^{-1} v_{in} - A_1 z^{-1} v_{out}\right).$$

In SystemC we can implement such difference equation by a process, which is activated every T seconds and which computes the output $y(t)$ from the input $x(t)$ as follows:

```
//straigthforward implementation of a IIR-filter
//for a more efficient implementation see, e.g., [Lyons, 1997]
          ...
for(unsigned long i=1;i<zyn;i++) zy[i]=zy[i-1];
for(unsigned long i=1;i<zxn;i++) zx[i]=zx[i-1];
zx[0]=x;
y=0.0;
for(unsigned long i=0; i<zxn; i++) out+=B[i]*zx[i];
for(unsigned long i=1; i<zyn; i++) out-=A[i]*zy[i];
zy[0]=y;
y\=A[0]; ...
```

Modeling of simple linear networks.   For many applications it is often very important to include simple linear networks into the system level model. For example, an equivalent circuit of a subscriber has to be included for system level simulations of subscriber line drivers in telecom applications. Such networks can be switched, e.g. a subscriber from 'on hook' to 'off hook'.

To model such behavior in SystemC, the network has to be transformed into analog equations. Therefore the input and output voltages and currents have to be defined. The use of a state space representation has the advantage that we can easily access and/or modify the physical quantities such as the magnetic flux or the charge.

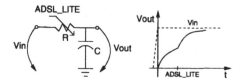

Figure 10.4.   Simple switched network

The network shown in figure 10.4 has the following state space representation in general, with x as an internal state:

$$\dot{x} = Ax + Bu,$$
$$y = Cx + Du.$$

In order to obtain a SystemC model we must determine the parameters A, B, C, D. Then we can apply one of the known methods for solving the differential equation:

- First, we assign the charge q to the state variable s: $s = q = Cv_{out}$;

- Second, we set up the system of equations using the equation $I = \dot{q} = (v_{in} - v_{out})/R$. Then we obtain:

$$\dot{q} = -\frac{1}{RC} q + \frac{1}{R} v_{in},$$

$$v_{out} = \frac{1}{C} q + 0 \ v_{in};$$

- If we have a function SS(...) which solves the equations we can model the switched network of figure 10.4 as follows:

```
SC_MODULE(sw_rc) {
    sca_quantity_in   input;
    sca_quantity_out  output;
    sc_in<bool> ADSL_LITE; //control in-port
        :
    void  sig_proc() {
        double tmp;
        A1[0]=1.0/(C*R1);       A2[0]=1.0/(C*R2);
        B1[0]=-1.0/R1;          B2[0]=-1.0/R2;
        C1[0]= 1.0/C;           C2[0]= 1.0/C;
        D1[0]= 0.0;             D2[0]= 0.0;
        id1.T=T;                id2.T=T;
        while(true) {
            //predefined function for state space
            //state vector S will be hold
            if(ADSL_LITE) tmp= SS(A1,B1,C1,D1,S,id1,input.read());
            else          tmp= SS(A2,B2,C2,D2,S,id2,input.read());
            output.write(tmp);
        } }
            ...             };
```

10.3.2   Stiff and Reactive Systems

In stiff and reactive systems constant time steps are not applicable:

- Stiff systems combine very large and very small time constants. For the application of constant step widths we would have to choose a sampling frequency according to the lowest time constant. However, this leads to unacceptable simulation times. Usually the best way to handle stiff systems is to simulate them by powerful DAE solvers like DASSL [Petzold, 1983] or by coupling with a commercial analog simulator (see section 4);

- Reactive systems are discrete processes which 'react' on conditions that can be statisfied at arbitrary points in time. Such conditions lead to synchronization between the discrete processes and analog modules at non-equidistant points in time: either very often or after a long period of inactivity.

In the remainder of this section we will discuss some typical models for reactive subsystems. Often reactive control systems model a discrete controller which controls an analog environment [Grimm et al., 2002]. Then we can make the following simplifications:

- We can model the analog environment by a linear differential equation or a transfer function $H(s)$;

- The communication between discrete controller and analog environment only occurs on discrete events coming from the discrete controller.

For linear differential equations and transfer functions we can compute an explicit solution. The explicit solution avoids additional numerical integration and allows us to compute the actual output value whenever the output signal of the analog module is required. If the communication is only initiated by the discrete controller we do not have to propagate events from the analog domain to the discrete simulator. This allows us to model the analog environment as a pure 'slave' simulator, which is only called when its outputs are required or when there are new inputs for the analog 'simulator'.

Simple Synchronization. If we have *one* analog simulator which is coupled with a discrete event simulator such as a SystemC simulator, we must synchronize the simulators at least under the following conditions:

- After a discrete event input signal of an analog model has changed its value new states of the analog system must be computed;

- The analog simulator must compute the output of a analog model at the points in time at which the digital controller reads the analog outputs.

SystemC provides very efficient means that allow us to implement such synchronization in a very direct way: the master/slave library, which maps port-accesses to remote method calls. As a very simple example the model of a continuous time integrator is given. For such a simple analog system we can compute the output in a direct way for arbitrary step widths. We assume a piecewise constant input from a discrete system $u(t)$, which has been constant from t_{last} until the actual point in time t_{now}. Given the output and state $y(t_{last})$, we can compute the output $y(t_{now})$ and the $state(t_{now})$ by:

$$state(t_{now}) = y(t_{now}) = u(t_{last}) * (t_{now} - t_{last})$$

In SystemC we can model such an integrator as follows:

```
SC_MODULE(integ) {
  sc_inslave<double> u;
  sc_outslave<double> y;
  double state;
  sc_time t_last;
      virtual void sig_proc() {
          state += y_last*(now()-t_last);
    y_last=u.read(); t_last=now();
          y.write(state);
      };
  SC_CTOR(integ){
    SC_SLAVE(sig_proc,u); // state transition function
    SC_SLAVE(sig_proc,y); // output function
  }
};
```

In the above model a function sig_proc is called whenever an input signal is written. The same function is also called when the output signal y is read. The function sig_proc computes an output and a new state.

A model of a transfer function $H(s)$.　　Signal processing systems and control systems on system level are specified by static nonlinear (combinational) and dynamic linear equations. The simulation of static nonlinear functions can also be done directly in a discrete event environment. Dynamic linear equations are usually described by transfer functions $H(s)$:

$$H(s) = \frac{a_m s^m + \ldots + a_1 s + a_0}{b_n s^n + \ldots + b_1 s + b_0}.$$

If $m < n$ we can find a partial fraction representation using for example Matlab:

$$H(s) = \frac{r_1}{s - p_1} + \frac{r_2}{s - p_2} + \ldots + \frac{r_n}{s - p_n}.$$

From the partial fraction expansion we can compute the impulse response. Provided there are no multiple poles, we obtain:

$$h(t) = r_1 e^{jp_1 t} + r_2 e^{jp_2 t} + \ldots + r_n e^{jp_n t}. \tag{10.3}$$

Equation 10.3 describes the output of a transfer function with a Dirac impulse $\delta(t)$ as input. However, discrete event signals are piecewise constant. Each constant signal segment is the integration of a Dirac impulse over time. The integration and the previous values are modeled by a complex variable `state`. As a SystemC module we obtain:

```
...
vector<double> a;
vector<complex<double> > pole;
vector<complex<double> > state;
sc_time last_change;
void sca_spartial::sig_proc(){
  double output=0.0;
  sc_time now = simcontext()->time_stamp();
  sc_time t = now - last_change;
  last_change = now;
  for (unsigned int i=0; i++ < a.size();)
  { state[i] = exp(t.to_seconds()*pole[i])
                 *(state[i]-a[i]*u.read())
                 +a[i]*u.read();
    output += state[i].real();
  }
  y.write(output);
}
...
```

Control blocks. In control systems dynamic linear controllers are also specified by parameters of P, PI, and PID controllers. Their parameters k_P, T_I and T_D have a direct meaning in the time domain. For example, in SystemC a PI controller with the parameters T_I and k_P can be modeled as follows:

```
void asc_s_pi_controller::sig_proc()
{       sc_time now=simcontext()->time_stamp();
  sc_time t = now-last_change;
  last_change = now;
  state += x.read()*t.to_seconds();
  y.write( k * (T*state + x.read() ) );
}
```

A description of problem specific simulation techniques for power electronic application is given in [Grimm et al., 2002].

10.3.3   RF Wireless Applications

An efficient simulation of Radio-Frequency wireless applications such as UMTS, wireless LAN, and Bluetooth in a direct way is often impossible. The problems are the very high frequencies in wireless communication systems. For example, for HYPERLAN 5GHz are required for a passband simulation in which the signal bandwidth is only 20 MHz. The high passband frequency leads to a very high sampling rate in the system simulation. The consequence is a very low simulation performance. Because the realization in SystemC is very complex we only give a short overview of the underlying methods that permit the simulation of such systems.

A RF system like that shown in figure 10.5 consists mainly of signal processing modules. Only the front end block (RF Rx) includes RF components. However, those few modules will dominate the simulation performance of a 'classical' time domain simulation, owing to the high base band frequency. The following modules working on the orders lower pass band frequency.

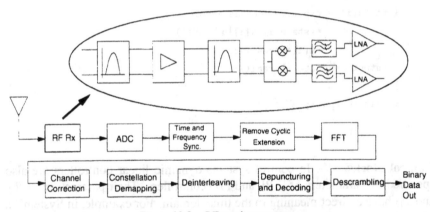

Figure 10.5.   RF receiver system

A significant acceleration of the simulation is possible when we consider the following properties of RF systems: digital modulation techniques use the magnitude r and the phase ϕ of a carrier signal to transmit information. Therefore the information transmitted is independent of the carrier frequency itself. The idea of *base band simulation* is to use a carrier frequency of zero. The required sampling rate then depends only on the modulation signal bandwidth, and is independent of the carrier frequency. Figure 10.6 shows the effect of such transformation in the frequency domain.

Formally the modulated carrier signal $x(t)$ can be described as:

$$
\begin{aligned}
x(t) &= r(t)\cos(2\pi f_c t + \phi(t)) \\
&= Re[r(t)e^{j(2\pi f_c t + \phi(t))}] = Re[r(t)e^{j\phi(t)}e^{j2\pi f_c t}],
\end{aligned}
$$

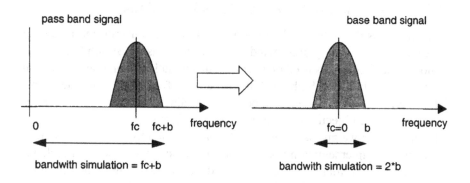

Figure 10.6. Pass band and base band representation of signals

where $r(t)$ is the modulating signal, $\phi(t)$ the modulation phase and f_c the carrier frequency. The term which includes the carrier frequency f_c can be separated from the signal part which contains the transmitted information. The signal that contains the information is independent of the carrier frequency f_c:

$$v(t) = r(t)e^{j\phi(t)}.$$

This signal is called the *complex low pass equivalent*, or the complex envelope. For the base band signal the carrier frequency f_c is converted to zero. With $s_i = r\cos(\phi)$ and $s_q(t) = r\sin(\phi)$, we obtain

$$v(t) = s_i(t) + js_q(t).$$

The resulting base band signal is represented by two signal parts: $s_i(t)$ and $s_q(t)$. The amplitude and phase of the carrier signal can be computed from these signals at each point in time.

Like the signals, the models of RF blocks must also be transformed from the pass band to the complex base band representation, e.g. a band pass filter in pass band simulation corresponds to a low pass in base band simulation. The equivalent low pass processing can be defined more formally using the Hilbert transform [Jeruchim et al., 1992].

For modeling such modules in SystemC, are the methodologies described for signal processing systems can be applied. Instead of the data type `double`, the C++ Standard Template Library (STL) type `complex<double>` will be used. The sampling frequency must then correspond to the pass band and not to the base band frequency.

The main disadvantage of this simulation approach is that effects outside the signal bandwidth cannot be represented in the base band signal, e.g. harmonics of the carrier frequency. Unfortunately such effects could have an impact on the performance of further receiver components.

10.4   Overview of SystemC-AMS Extensions

The methods described in the previous sections demonstrate the modeling and simulation of analog and mixed-signal systems using SystemC 2.0. All methods described are focused on the efficient simulation of analog modules in a system simulation on a high level of abstraction. To achieve a sufficient simulation performance different methods for modeling and simulation are used, depending on the objectives of the simulation problem.

In this section we give an overview of *concepts* for a SystemC-AMS extension from the implementation point of view [Einwich et al., 2002b], [Vachoux et al., 2003b]. Similar ideas are discussed in the on going standardization process. We show that although the previously discussed methods for modeling, simulation, and synchronization are application and problem specific, SystemC-AMS is a general and generic framework for system level simulation and supports a seamless design methodology from system level down to circuit level. Note that the focus of SystemC-AMS is not to add an analog simulator to SystemC. Instead, SystemC-AMS provides a framework for the system level simulation of complex mixed-signal systems with open interfaces that allow us to integrate appropriate simulation methods as well as commercial simulators such as SABER, for example.

The interfaces and methods provided by SystemC-AMS are structured into different layers [Einwich et al., 2002a]. Figure 10.7 gives an overview of these layers.

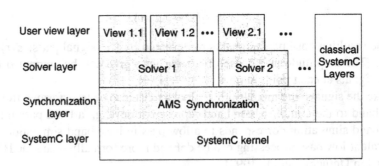

Figure 10.7.   Layers of SystemC-AMS

The *view layer* transforms the user description into a system of equations. For example, Modified Nodal Analysis (MNA) is applied to an electrical network to create matrices which represent the system of equations.

The *solver layer* provides different analog solvers, e.g., for solving DAEs or computing transfer functions $H(s)$. Solvers are instantiated from the view layer. A solver is able to solve a system of equations provided from certain view layers.

The *synchronization layer* exchanges signal values between analog solvers and the SystemC discrete event kernel. Furthermore, it determines the next simulation time step. The synchronization layer encapsulates analog solver instances from each other and the analog solvers from the discrete event kernel.

On all layers the interfaces provided are generic and are realized by simple, yet efficient methods. This allows us to add problem or application specific methods, as discussed, for example, in the previous sections, or to integrate existing simulators. The `sca_` prefix will be used for all SystemC-AMS extensions to prevent symbol duplications. The new simulation methods can be added to an analog base module declared by the macro SCA_MODULE by (multiple) inheritance. This macro hides the inheritance from a `sca_module` - base class. The class `sca_module` provides functionality which is common for all analog modules and is derived form the standard SystemC `sc_module` class.

In the following, examples for the definition of application specific analog modeling styles will be given. Using SystemC-AMS extension classes a static data flow modeling style (sdf), which is influenced by the data flow modeling provided by tools like COSSAP or SPW, can be provided as follows:

```
SCA_MODULE(sdf_module),   sca_sdf_synchronization,
                          sca_linear_solver, sca_ac_solver
                          sca_behavior_view
{
  virtual void init(){};      // initialization method
  virtual void sig_proc();     // simulation method
  virtual void post_proc(){};  // post processing method
  SCA_CTOR(sdf_module)
  {
  //registration of data-flow methods
  //provided by sca_sdf_synchronization
    SCA_SDF_INIT(init);          //initialization method
    SCA_SDF_RUN(sig_proc);       //signal processing method
    SCA_SDF_POST(post_proc);     //post processing methods
  }
};
//type definition for module ports using interfaces which are
//implemented by a sca_sdf_synchronization channel
typedef sca_port<T,IF_in_sca_sdf_synchronization>  sdf_in<T>;
typedef sca_port<T,IF_out_sca_sdf_synchronization> sdf_out<T>;
```

The class `sca_sdf_synchronization` signs the module to be synchronized with other modules by a static data flow scheduler. This base class provides also an interface for the scheduler (e.g., the macros for method registration like SCA_SDF_RUN). The classes sca_linear_solver, sca_ac_solver register available solvers and providing an interface for the solvers. In the above example a linear time domain solver and a small signal frequency domain (AC-solver) will be available. The solver interfaces provided will be used by the view layer. A behavior view layer will provide methods for the description of transfer functions or state space systems. Thus the sdf_module class provides similar modeling facilities like described in sections 2.2 and 3.1 — however, in a more comfortable way.

The following example illustrates the model of a simple low-pass filter using the modeling style described above. The optional frequency domain implementation will be used by the AC solver which calculates the overall frequency domain behavior using this implementation and the structure of the system which will be accessed via the sca_module class interfaces.

```
struct low_pass: sdf_module
{
  sdf_in<double>  in;   //port declaration
  sdf_out<double> out;

  double fc;            //parameter for cut-off frequency
  void init() {
          B[0]=1.0;     //Coefficients of transfer function
          A[0]=1.0;
          A[1]=1.0/(2.0*M_PI*fc) }

  void sig_proc() {
          //transfer function method provided by sca_behavior_view
          out= sca_ltf(A,B,S,id,in);
                }

  void ac_domain() {
    //complex transfer function for AC simulation
    //SCA_AC macros allow access to complex global system of equations
    //SCA_W returns the simulated frequency
    complex<double> j(0,1);
    SCA_AC(out)=B[0]/(A[0]+j* SCA_W*A[1]) * SCA_AC(in);
                }
  SCA_CTOR(low_pass) {
      //registration frequency domain implementation
      SCA_AC_DOMAIN(ac_domain);  }
```

```
private:
  sca_vector A, B, S; //vector class holds coefficients and states
  sca_dae_id id;     //signs system of equations
};                   //if more than one used per module
```

The next example demonstrates the definition of a module base class which supports the description of conservative linear electrical networks. For conservative systems a global system of equations has to be set up and solved. Every conservative component has to contribute to the global system of equations. Thus the conservative view (sca_conservative_mna_view) must provide methods for the description of the system of equations contribution, must set up one system of equations for connected components, must instantiate one solver for this system of equations and must embed this solver in the defined synchronization domain.

```
SCA_MODULE(linear_electric_component), sca_sdf_synchronization,
                                       sca_linear_solver,
                                       sca_conservative_mna_view
{               :
  virtual void matrix_stamps();    //system of equations contributions
                :                  //for a Modified Nodal Analysis (MNA)
  SCA_CTOR(linear_electric_component)
  {
  //registration of method for matrix stamps
  //provided by sca_conservative_view
  SCA_MATRIX_STAMPS(matrix_stamps);  } };
```

A resistor, for example, would modify the matrix stamps of the system of equations as follows:

```
class sca_r :  public linear_electric_component {
  sca_node a; //conservative ports
  sca_node b;
  public:                    :
    //matrix stamps for a resistor, will be used to set up
    //the system of equations for the Modified Nodal Analysis (MNA)
    void matrix_stamps() {
          sca_a[a,a]+= 1.0/R; //sca_a global matrix for the system
          sca_a[a,b]+=-1.0/R; //of equations which is provided by
          sca_a[b,a]+=-1.0/R; //sca_conservative_mna_view
          sca_a[b,b]+= 1.0/R; }
        :
  SCA_CTOR(sca_r) { } };
```

In a similar way we can define voltage sources (sca_v), capacitors (sca_c) and many other electrical or non-electrical components. Such components can also contain ports which can be connected with channels of other domains such as the sca_vsdf source in the following example. This source applies a voltage to the network with a value derived from the data flow in port. Electrical components can be connected in a hierarchical module in the same way as standard SystemC modules or similar to SPICE or VHDL-AMS netlists:

```
SC_MODULE(rc_netlist) {      //standard SystemC-module declaration
    sca_sdf_in<double>  in;   //data-flow in-port used in voltage source
    sca_node         out;    //electrical port
                       :
    SC_CTOR(rc_netlist) { //SystemC-constructor
        v1= new sca_vsdf (''v1'');   //voltage source
            v1->value(in);          //the applied voltage is derived
            v1->a(w1);              //from a data-flow port
            v1->b(gnd);
        r1= new sca_r(''r1'');
            r1->value=100.0;
            r1->a(w1);
            r1->b(out);
        c1= new sca_c(''c1'');
            c1->value=1e-6;
            c1->a(out);
            c1->b(gnd); }};
```

Thus components of different analog domains and standard SystemC modules can be instantiated and connected in hierarchical modules. In order to connect different modeling styles converters can be instantiated either in an explicit way, or implicitly by general purpose signals which realize a number of different interfaces. Examples for such signals and their application in a *refine and validate* design flow are discussed in chapter 11.

10.5   Conclusions

At system level a very high simulation performance is required. The required performance can be achieved by the combination of different problem and application specific simulation techniques.

In section 3 we have discussed a number of simple but efficient methods that allow us the simulation of behavioral models, or executable specifications of analog systems. Most of these methods also work with SystemC 2.0. In section 4, we have shown possible extensions that allow us to integrate different simulation methods to a simulation framework: SystemC-AMS. A first proto-type implementation of the proposed SystemC extensions has been described.

Although analog extensions of SystemC allow designers to simulate some analog systems, they are not a replacement for mature circuit simulators such as SPICE or SABER. SystemC-AMS provides us with an overall framework. In such an overall framework we can combine different methods for the simulation of analog and mixed-signal systems. Therefore the SystemC-AMS framework has not only the potential to speed up system simulation. SystemC-AMS can also be seen as a system level integration platform, which allows us to deal with the increasing heterogeneity on system level — both in modeling formalisms and in tools.

Acknowledgments

The authors thank the colleagues of Infineon Technologies MDCA Villach, especially Gerhard Noessing and Herbert Zojer, for supporting the activities of AMS modeling with SystemC and the numerous practical inputs and hints. Furthermore, the authors thank Uwe Knoechel from the Fraunhofer Institute IIS for contributing to the RF modeling part. We like to thank especially the reviewers for their criticism and the valuable hints for the improvement of the paper.

Chapter 11

Modeling and Refinement of Mixed-Signal Systems with SystemC

Christoph Grimm

University of Frankfurt, Germany

Abstract   This chapter describes methods for the simulation and the design of complex signal processing systems with SystemC. The design starts with a block diagram which can be simulated in SystemC. Refinement steps transform the block diagram to a more detailed description. In the detailed description the execution of functional blocks is controlled by methods which explicitly determine step widths. Furthermore, ports, for example clock and enable ports, are added. The refinement permits the efficient interactive exploration of the design space of mixed-signal systems.

11.1   Introduction

Sophisticated tools for the synthesis at the RT and behavioral level permit a *specify and synthesize* design flow of digital systems. Together with the re-use of complex IP blocks this allows designers to keep up with the increasing complexity of digital hardware. However, the design of heterogeneous systems is far more complex and tools are less mature. Therefore the design of such systems is more interactive, and languages and associated methodologies for modeling and design are more important.

SystemC, and even more SpecC [Gajski et al., 1997] support the interactive design of hardware/software systems by a *refine and evaluate* methodology [Gajski et al., 1994a]. An untimed functional model is modified by 'small' design steps. These design steps successively add run times of a hardware or software realization to untimed processes and change the semantics from a purely sequential execution to a partially concurrent execution. Although the design steps require only small modifications in the model, their effect on the system is tremendous: Components are shifted from software to hardware, or

W. Müller et al. (eds.),
SystemC: Methodologies and Applications, 299–323.

vice versa; without having to design all the necessary interfaces. This allows designers the efficient, interactive design of digital hardware/software systems.

However, many systems include a variety of different analog components [Vachoux et al., 2003b]: Control systems in the automotive domain, or line drivers or RF frontends in the telecom domain, for example, are typical applications. Their realization includes software running on DSPs, dedicated hardware for digital signal processing, sample rate converters, PLLs, A/D converters, analog filtering and signal conditioning, even power electronics and mechanical components. Unfortunately, up to now the refine and evaluate approach seems to be restricted to the design of hardware/software systems.

Related Work.   VHDL-AMS [IEEE, 1999a] permits the modeling of mixed-signal modules, such as PLLs or A/D converters at different levels of abstraction. Unfortunately there is no specific support for the modeling of communication: Analog parts are modeled using (continuous) quantities, digital components communicate via (discrete event) signals. Whenever analog and digital components communicate a converter has to be specified in detail. Therefore even small modifications of the system's partitioning result in significant, time consuming modifications of the model.

Simulink [MathWorks, 1993] is very common for the interactive design at system level. Designers can interactively create, simulate, and modify block diagrams. Unlike in VHDL, there is no strict type checking. However, a general support for different models of computation which is necessary for modeling heterogeneous systems at different levels of abstraction is missing. Therefore Simulink is restricted to the evaluation of abstract models.

SPW, Cossap, and Ptolemy/HP permit the efficient simulation of complex signal processing systems using the dataflow model of computation. Ptolemy II [Liu et al., 1999] in addition supports the interoperation of CAD tools/simulators that have different underlying models of computations: discrete event systems; continuous time systems; static data flow; finite state machines; communication sequential processes; process networks; etc..

This chapter presents methods for the simulation and design of complex heterogeneous signal processing systems in SystemC. Many methods for the simulation are similar to those found in Cossap or Ptolemy II. Compared to these tools, the methods introduced in this chapter specifically support an interactive refine and evaluate methodology. This methodology covers the tasks:

- Partitioning analog/digital/software;

- Determination of step widths and quantization;

- Binding of communication to physical signals.

Section 2 of this chapter introduces basic methods for the simulation of signal processing systems. Section 3 gives hints and examples for the implementation of these methods in SystemC. Most of the source code is taken from the ASC library [Grimm et al., 2003], an early prototype of SystemC-AMS [Einwich et al., 2002b, Vachoux et al., 2003b, Vachoux et al., 2003a], and is simplified significantly in order to illustrate the basic methods and concepts. Section 4 describes the use of SystemC for the refinement of an executable specification to different architectures of signal processing applications. Section 5 gives an overview of a case study.

11.2 Basic Methods and Definitions

Block Diagrams. At system level, signal processing systems are modeled by block diagrams. A block diagram consists of blocks and directed signals between the blocks. The blocks perform different functions and are specified using different models of computation. Signal processing functions map a space of n input signals $in_i \in V_{in,i} \times T_{in,i}$ to a space of m output signals $out_j \in V_{out,j} \times T_{out,j}$:

$$sigproc \in in_1 \times \ldots \times in_n \rightarrow out_1 \times \ldots \times out_m$$

The value ranges V of the signals are either the real numbers or discrete approximations. The time bases T of the signals are the real numbers modeling continuous time or discrete approximations of continuous time. Signal processing functions are specified by differential and algebraic equations (DAEs) or difference equations in general. At system level most notably linear dynamic and nonlinear static functions are used for the specification of signal processing functions. We call blocks with signal processing functions *signal processing blocks*. Contiguous networks of signal processing blocks are called a (signal processing) *cluster*. Signal processing blocks are connected via directed edges that connect output signals with input signals. The meaning of a connection from an output signal out_j to an input signal in_i is a mathematical equation, for which we use the operator '==', and not an assignment (operators '=', ':='):

$$out_j(t) == in_i(t) \qquad \forall t \in T.$$

The function of the signal processing cluster is determined by the equations that specify the blocks, and by the equations introduced by the structure of the block diagram. Note that we also call the connection a signal.

Figure 11.1 shows a block diagram which consists of four blocks. Three of them form a signal processing cluster, and one performs other functions. For the simulation of block diagrams, all blocks of a signal processing cluster are assigned to a *coordinator* (see figure 11.1). A coordinator performs two tasks:

1 Simulation of a signal processing cluster;

302

2 Synchronization of the signal processing cluster with the discrete event environment.

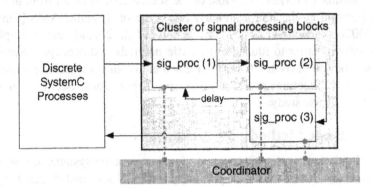

Figure 11.1. A block diagram of one block with SystemC processes, and a cluster of three signal processing blocks. A coordinator controls the simulation of the signal processing cluster.

Simulation. For the simulation of a cluster of signal processing blocks we can apply the following methods:

- *Set up equations for the cluster:* The equations which specify the function of the signal processing blocks and the equations introduced by the connections specify a set of DAEs. The output signals of the signal processing cluster are then computed by a single simulator that solves this system of equations;

- *Relaxation:* Cyclic dependencies in the block diagram are broken into acyclic structures by introducing a sufficiently small delay of T seconds, as shown in figure 11.1. (In general this delay can also be zero. Then we say that the simulator *iterates*. In the following we assume a delay > 0). The resulting acyclic structure is then ordered the direction of the signal's flow. After that the outputs can be computed successively from already know inputs by simulating each block for his own (static dataflow). When all blocks are computed the cluster waits for T seconds. Note that the delay determines the frequency with which the signals between the signal processing blocks are sampled. In figure 11.1, for a given input of the SystemC kernel and a known delayed value, the coordinator would first ask block *sigproc1* to compute his outputs, then *sigproc2*, and finally *sigproc3*. After a delay the same procedure would start again, and so on.

The first approach is used, for example, in VHDL-AMS simulators. It is a very precise and reliable method. However, setting up equations for a whole cluster requires that all blocks in a cluster are modeled by DAEs, procedural assignments, or explicit equations, and that all equations are solved by one 'solver'. Therefore the combination of different methods for the simulation of analog systems is not supported by this method.

The second method is used in Simulink, SPW, Cossap, and also in AMS extensions of SystemC. It permits the very efficient simulation of heterogeneous signal processing systems, because the static data flow model of computation is used for computing the outputs of the cluster. Because every block is simulated independently from each other, this approach allows designers to mix different models of computation in a cluster. A drawback of this method is its restricted convergence. Relaxation should only be applied to simulate weakly coupled systems. This requirement is usually met by block diagrams used at system level. Table 11.1 compares the methods for the simulation of block diagrams.

Table 11.1. Simulation of signal processing clusters

Set up equations	Relaxation
precise	convergence unsure
also conservative systems	only weakly coupled systems
homogeneous systems	heterogeneous systems
slow	very fast

Signals and Synchronization. The signals of the signal processing clusters are fundamentally different from discrete event signals. Discrete event signals change their values at discrete points in time and are constant between these points in time. The signals of signal processing clusters are discrete approximations of continuous time signals (see figure 11.2). For simulation they are represented by a sequence of discrete samples. However, in signal processing clusters values between the samples are interpolated (in figure 11.2 between last_val and now_val, for example). The quantization and interpolation introduces an error to the simulation (*quantization error*). Therefore the sampling rates must be much higher than twice the maximal frequency, so that the signals do not change their values significantly between two sample points.

Signals that are written from a discrete process are piecewise constant from the last value written at a point in time to the actual point in time. Digital, discrete event signals do not approximate a continuous time signal — therefore they have no quantization error at all. However, the value at the actual point in time can be changed in delta cycles of the discrete event simulator. Therefore the final value is known only once all delta cycles have been exhausted.

When discrete event blocks are connected with signal processing functions via a signal, this signal must have *both* properties of a continuous time signal and

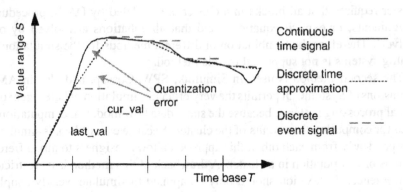

Figure 11.2. Continuous time signal, discrete time approximation and a discrete event signal

of a discrete time signal. Furthermore, we must ensure that no deadlocks occur, for example if the discrete processes need inputs from the signal processing cluster and vice versa at the same time. Figure 11.3 shows the potential problem

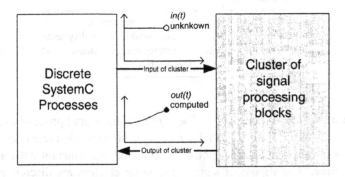

Figure 11.3. A simple synchronization between signal processing cluster and discrete event processes

caused by cyclic dependencies between a signal processing cluster and discrete processes. A simple, precise, predictable, and efficient synchronization works as follows:

At a point in simulated time t, when the signal processing cluster is scheduled to compute a new value $out(t)$ of the output signal the actual input value $in(t)$ is still unknown, because the output can set an event which would modify its input in the next delta cycle. Then the signal processing cluster would have to reset its internal states and compute the output again. Unfortunately this is complex, time consuming, and, in a heterogeneous tool environment, not supported by all blocks, respectively, their simulators.

In order to have a predictable behavior of the simulation a signal processing block that reads a discrete event signal gets the value of this signal before it has been modified at t, that is, its value before or at the first delta cycle of the discrete event simulation $(t-)$. Then the signal processing cluster has a known input from the last change untill the actual point in simulated time t and can compute output values including $out(t)$. This synchronization scheme cannot have deadlocks, because the signal processing cluster has always known inputs from the past and computes all necessary inputs for the discrete processes at the actual simulated time.

Together with the relaxation based computation of the cluster, which is implemented using a static data flow model of computation, we get the following overall algorithm for the simulation of block diagrams:

> **simulation_of_cluster**
> Before simulation:
> - Break up cyclic dependencies by insertion of a delay;
> - Sort all blocks in signal flow direction.
> During simulation:
> Every Δt do:
> - Read input values $in(t-)$ from DE signals;
> - Compute signal processing blocks in determined order;
> - Write DE output values $out(t)$ to DE signals;
> - Simulate all DE delta cycles at t.

The above method can be implemented easily in SystemC. In the following section we describe some details for the implementation of this method in SystemC.

11.3   Implementation in SystemC 2.0

The code examples given in the following are based on the ASC library (Analog SystemC library, [Grimm et al., 2003]), an early prototype of SystemC-AMS. The ASC library supports simulation of signal processing systems, but its main focus is the validation of the refine and evaluate design methodology. In order to reduce the complexity and to better illustrate the methods applied, we have simplified and omitted many details of the implementation. For example, we do not explain how to realize a static data flow scheduler, but we explain how such a scheduler can be integrated in SystemC. Furthermore, we have reduced the complexity of the interfaces to a small number of functions, which is sufficient for a basic functionality.

In order to simulate signal processing clusters in SystemC with the methods discussed in section 2, we need the following classes:

- Classes that allow us the simulation of signal processing blocks of a cluster. Such classes are desribed in chapter 10;

- A class that realizes the coordinator of a cluster;

- A signal class that implements the synchronization scheme explained above, together with the required ports and interfaces.

In the following, we give some hints for the implementation of these classes.

A Signal Processing Module Class.   We assume that elementary signal processing functions such as transfer functions are encapsulated in objects of the class `asc_module` that is inherited from the class `sc_module`:

```
class asc_module: sc_module {
  asc_module(const char* nm, asc_coordinator* c): sc_module(nm) {...}
  virtual void sig_proc() = 0;
  ...
  asc_coordinator* coordinator;
}
```

Signal processing modules have a virtual method `sig_proc` that is called by the coordinator. This method reads input signals and computes output signals for a given time step. In order to support a further refinement this method must support variable time steps, of course. Examples for such methods are given in chapter 10 or in [Grimm et al., 2002].

A *Generic* Coordinator Class.   The functions of a coordinator can be implemented in a module of the class `asc_coordinator`. During elaboration signal processing modules can be assigned to a cluster that is managed by this coordinator. In the coordinator a vector `block_list` contains pointers to all modules of the cluster. An event `synchronization` is used for the synchronization with the discrete event kernel of SystemC. This event triggers the execution of the method `simulate` which simulates the blocks assigned to this cluster. The method `request_sync` allows blocks or signals to request a synchronization in a future point in time. `register_block` adds a new block to the cluster controlled by this coordinator. In SystemC 2.0:

```
struct asc_coordinator: sc_module {
  protected:
    vector<asc_module*>  block_list;      // blocks in cluster
    vector<asc_signal*>  signal_list;     // signals in cluster
    sc_event             synchronization; // event for sync
    sc_time              ts;              // step size
  public:
    SC_HAS_PROCESS(asc_coordinator);
    asc_coordinator(): sc_module(sc_module_name("coordinator"))
    { SC_METHOD(simulate) sensitive << synchronization; };
    virtual void end_of_elaboration();
    virtual void simulate();
```

```
virtual void request_sync(const sc_time& t, asc_module* m);
void register_block(asc_module* m);
void register_signal(asc_signal* s);
};
```

The methods simulate, end_of_elaboration, and request_sync are *virtual*. Therefore they specify an interface which can also be realized by another coordinator that realizes different methods for the simulation and synchronization of the signal processing blocks. For the simulation of block diagrams as discussed in section 11.2, the method simulate computes all signal processing functions and notifies the event synchronization for the next cycle in ts time units:

```
void asc_coordinator::simulate() {
  for (int i=0; i < block_list.size(); i++)
    block_list[i]->sig_proc();
  for (int i=0; i < signal_list.size(); i++)
    signal_list[i]->update();
  synchronization->notify(ts);
}
```

The implementation of the other methods is straightforward. A static data flow scheduler can be implemented in the method end_of_elaboration. For smaller projects, this can also be omitted. In that case a designer must register the blocks manually in the signal flow order.

A Signal Class for *Interactive* Modeling. In order to support the interactive modeling of signal processing systems we introduce a signal type that realizes both the interface of the discrete event signals, and an interface used for coupling analog blocks — the methods read_a and write_a. This allows designers to connect such a signal to ports of the discrete modules and to ports of signal processing blocks.

```
template <class T>
class asc_signal_inout_if: virtual public sc_signal_inout_if<T> {
public:
  virtual const T& read_a() const = 0;
  virtual void write_a(const T&) = 0;
  ....
};
```

The binding of read/write functions to the appropriate communication methods of one of the two interfaces can be handled in the ports: The ports sc_in and sc_out which come with SystemC 2.0 always call methods of the interface of discrete signals. We define a port class asc_inout for use in signal processing clusters; ports of the types asc_in and asc_out are subtypes of this port.

This port always calls methods of the interface of the continuous-time signal interface:

```
template <class T> const T& asc_inout<T>::read() const {
  return (*this)->read_a();
}
template <class T> void asc_inout<T>::write( const T& value_ ) {
  (*this)->write_a( value_ );
}
template <class T> const T& asc_inout<T>::read_a() const {
  return (*this)->read_a();
}
template <class T> void asc_inout<T>::write_a( const T& value_ ) {
  (*this)->write_a( value_ );
}
```

Note that most of the event handling is useless and could be omitted. If no events on continuous time signals are needed, even the inheritance from the primary channel class is not required and can be omitted. However, we keep inheritance from the primary channel class. This allows us to use the update routine for the synchronization between the signal processing cluster and the discrete kernel:

```
template <class T>
class asc_signal: public asc_signal_inout_if<T>,
                  public sc_prim_channel {
  public:
    virtual void write(const T& value);   // called by DE ports
    virtual void write_a(const T& value); // called by SP ports
    virtual const T& read() const;        // called by DE ports
    virtual const T& read_a() const;      // called by SP ports
    asc_signal(): sc_prim_channel(sc_gen_unique_name("asc_signal")),
                  m_now_val(T()), m_last_val(T()){};
    asc_signal(asc_coordinator* c):
                  sc_prim_channel(sc_gen_unique_name("asc_signal")),
                  m_now_val(T()), m_last_val(T())
    { c->register_signal(this); }
    virtual void update();
  protected:
    T m_now_val, m_last_val;
    asc_coordinator *coordinator;
};
```

Discrete modules read the signal via ports, which call the method read(). The following implementation just returns the current value of the signal:

```
template <class T> const T& asc_signal<T>::read() const {
  // coordinator->request_sync(0,0);
  return m_now_val;
```

```
};
```

Note that at this place we also can add a request to synchronize the signal processing cluster (that is, to compute its actual output!) before returning the current value. The same applies for signal processing modules and their ports with the difference that these ports always call the method read_a:

```
template <class T> const T& asc_signal<T>::read_a() const {
  return m_now_val;
};
```

Remember that writing signals in the signal processing cluster is done in signal flow direction which has been determined statically before simulation. Therefore update/evaluate cycles as in the discrete event domain are not required. Of course, very sophisticated methods are possible for modeling signals in signal processing clusters. Most notably, FIFOs can be used to allow blocks to process larger segments of signals, or to enable the modeling of a multi rate data flow system. We just set the new value to the value actually written. After the computation of the cluster the coordinator sets the last value to this value.

```
template <class T> void asc_signal<T>::write_a(const T& value) {
  m_now_val = value;
}
```

When discrete event signals of SystemC 2.0 read or write inputs of the signal processing cluster, the following write method waits for an update cycle. This allows the signal processing cluster to always access the value before the first update cycle as discussed in section 11.2:

```
template <class T> void asc_signal<T>::write(const T& value) {
  m_now_val = value;
  request_update();
  // coordinator->request_sync(now);
}
```

Note that after the discrete event part has written this signal we can request the coordinator to do a new computation of the signal processing cluster. In the update methods the new values are finally written to the current value (at this point in time the signal processing cluster is already simulated):

```
template <class T> void asc_signal<T>::update() {
  m_last_val = m_now_val;
}
```

The abovementioned classes permit the modeling and simulation of block diagrams. In the next section we discuss the application of these classes in an interactive design methodology for the refinement of signal processing systems.

11.4 Refinement of Signal Processing Systems

The existing refinement methodology introduced with SystemC 2.0 (and SpecC) is obviously not applicable for signal processing systems: Untimed functional models of signal processing systems simply do not exist. In the following we introduce a design methodology applicable for the refinement of signal processing systems.

We distinguish three levels of refinement: The design starts with the specification of a *mathematical model*. This mathematical model is given by a block diagram and can be simulated using the classes discussed in section 3 of this chapter. Because this model can be simulated, we also refer to it as an *executable specification*.

By the *refinement of computation* methods are added that define sample frequencies, bit widths, and an order of computation for the digital parts. We call the resulting model a *computationally accurate model*. In a computationally accurate model all computations made in blocks or modules are done at the same points in time or on the same signal types as in the realization.

By the *refinement of interfaces* the abstract, non-conservative signals are successively replaced by a model of a physical realization, for example, conservative analog signals or different discrete signals. Signals which have been used at the border between different clusters are replaced by appropriate signal converters: A/D converters; sample rate converters, for example. After refinement of signals we have a *pin accurate model*. Furthermore, new ports are added to blocks for the digital signal processing. The added ports allow discrete controllers to explicitly control of the execution (e.g. enable or clock signals). In a pin accurate model all modules have the same ports as in the realization.

A computational accurate and pin accurate model is the starting point for the design of analog blocks or the hardware/software co-design of discrete modules. Both circuit level design of analog blocks, and hardware/software co-design are out of the scope of this chapter. However, hardware/software co-design is well supported by SystemC. The circuit level design of analog blocks usually goes bottom up. Behavioral models of analog blocks can be created as described in chapter 10 and integrated into a system level simulation. Of course, the methods described in section 2 of this chapter are also applicable for the integration of a circuit level simulator such as SPICE.

Figure 11.4 illustrates the refinement process. Note that — in difference to the established specify/synthesize design flows — there are not clear and fixed levels of abstraction at which specifications and implementations become available as inputs or outputs of tools! The small refinement steps are applied interactively to single modules. The application of the refinement steps can be done in nearly arbitrary order or even in parallel by different designers (!).

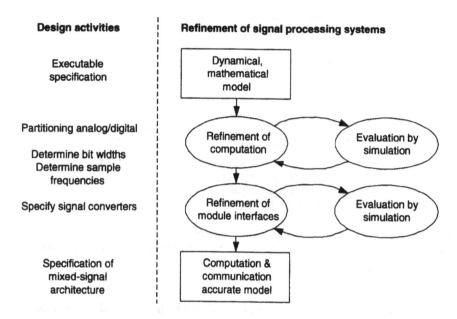

Figure 11.4.   Design activities in the refinement of signal processing systems

11.4.1   Specification by Mathematical Model

For the specification of the intended behavior an abstract description of the intended behavior is required. Usually this is a block diagram with continuous-time semantics. Therefore we also call it a dynamical mathematical model. The discretization comes usually with a selection of bit width and sample frequencies and is part of the design process. As explained in section 2 of this chapter, such a block diagram specifies an overall set of equations which can be solved by different methods. These methods are implemented in the coordinator. Because the coordinator is accessed via a set of virtual functions (which can also be abstract) these methods are generic.

Figure 11.5 illustrates the idea: Elementary functions, in the example the functions $sigproc_1, sigproc_2, sigproc_3$ are specified by transfer functions, for example. The connections have the meaning of additional equations. For a verification of this specification we compute a solution using the coordinator. The coordinator in this case just tries to simulate the ideal continuous time behavior of the system. In the case of the relaxation method without iterations (which is used by default for simulation) the coordinator can choose step widths for the sampling and the delay, so that the simulated behavior will be as close to the ideal behavior as possible.

With the classes defined in section 3 of this chapter and in chapter 10, the dynamical mathematical model can be created easily: Blocks with different

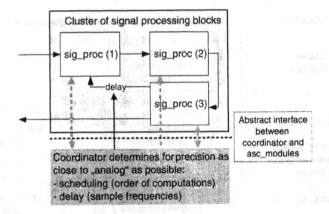

Figure 11.5. In the dynamical mathematical model the coordinator tries to approximate the continuous time behavior as precisely as possible by choosing delays as small as possible and executing all blocks in the signal flow's direction

dynamic linear or static nonlinear functions must be instantiated and connected using signals of the class asc_signal. This can be done in an interactive way by using pre-defined blocks such as PID controllers or transfer functions. Discrete event processes can as well access the signals of the class asc_signal directly without the need of specifying a converter. Therefore the design of the dynamical mathematical model which is used as executable specification is nearly as interactive as with Matlab. In comparison with Matlab, SystemC also permits the modeling of complex hardware/software systems as part of a signal processing system (or vice versa). This is very important, because in complex multimedia applications signal processing methods are very closely coupled with software with a complex control flow (e.g., motion detection in MPEG compression). Furthermore, SystemC is by far faster than Simulink because it is compiled, not interpreted.

11.4.2 Refinement of Computation

By the refinement of computation the blocks of the signal processing cluster are partitioned into blocks to be realized in digital electronic circuits and blocks to be realized by analog circuits. In order to evaluate different architectures it is sufficient to model the effects which come with a discrete time realization or an analog realization. These effects are most notably owed to the quantization, both in time and value, and owed to limitation effects for discrete time realizations. For analog realizations limitation and noise must also be considered. These effects can be added to a mathematical model designed with the classes of section 3 of this chapter with very little effort by inheritance or overloading of virtual functions.

For an *analog implementation* no further refinement is required - the realization is just a physical system whose behavior is 'analog' to the equations. Since limitation occurs frequently, we can also modify the method read a, assuming there are properties ub and lb modeling upper and lower bound for limitation:

```
template <class T> const T& asc_signal<T>::read_a() const {
  if (m_now_val > ub) return ub; else if (m_now_val < lb) return lb;
  return m_now_val;
};
```

The above modification permits one to parameterize analog signals to limit the values read by setting the properties lb, ub. Furthermore, different noise — for example, white Gaussian noise — can be added to the signals.

For a *digital implementation* (hardware or software) different sample rates have to be evaluated. If a cluster uses only one sample rate it can be modeled by setting the sample rate of the coordinator to the desired step width. This requires only modifying the parameterization of the coordinator and allows designers to get a first idea of appropriate sample rates. The modeling effort for changing the sample rate is very low, and the simulation is very efficient, because no details (clock signals, etc.) are added to the model. For modeling the different sample rates in multi-rate systems, the signal processing cluster can be split in smaller clusters. Each of the smaller clusters then has its own coordinator with its own sample rate — again, modeling effort is very low, and simulation speed remains very high. However, this also requires replacing signals by sample rate converters.

As soon as clock signals become available, a discrete controller can be written in a further refinement step. Instead of using processes sensitive to clock edges for evaluation of different rates, the interface between coordinator and the signal processing blocks should be used. The discrete controller divides a global clock signal. Whenever the controller wants to activate a signal processing block it calls the virtual function sig_proc of the module. Then the discrete controller replaces the coordinator and explicitly specifies a way in which to determine sample rates of blocks. This has the following advantages:

- The modeling effort is very low;

- The simulation speed is very high.

The quantization and the limitation of digital signals in the signal processing cluster can be modeled in the same way as in the analog signals. For the quantization we add a property resolution:

```
discrete_value=m_now_val-lb;           // subtracting offset
discrete_value/=(ub-lb);               // normalizing to 0 .. 1
discrete_value=floor(resolution*discrete_value);// quantization
discrete_value/=(resolution-1);        // back to value range 0 .. 1
```

```
discrete_value=discrete_value*(ub-lb)+lb;
```

Converters such as A/D converters, D/A converters, or sample rate converters usually have a complex, dynamic behavior. This cannot be modeled statically in a channel, because processes can only be used in modules. In order to introduce a signal converter without modification of the overall structure of the system hierarchical channels are used. In a hierarchical channel the dynamic behavior is modeled using discrete processes of a channel. Furthermore, the interface `asc_signal_inout_if` is inherited and realized:

```
class ad_converter: public sc_module, public asc_signal_inout_if {
  ad_converter(const char* name): sc_module(name) {
    // Declaration of processes
  }
  template <class T> const T& asc_signal<T>::read_a() const {
    // Method for reading from signal processing blocks
  }
  ...
}
```

The use of hierarchical channels allows designers to simply replace a signal by a converter with dynamic behavior.

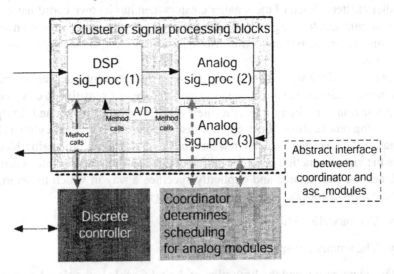

Figure 11.6. In the computation-accurate model, the design activities A/D partitioning and determination of sample rates have replaced the coordinator by a discrete controller, that determines them by means of discrete processes

Figure 11.6 shows a possible refinement of the dynamical mathematical model from figure 11.5: The signal processing function *sigproc* is realized in digital. Depending on clock signals already available a discrete controller calls the signal processing method via the coordinator interface; the signal processing function $sigproc_1$ therefore is no longer assigned to a coordinator. Because the functions $sigproc_2$, $sigproc_3$ are intended to be realized in analog, they are still assigned to the coordinator which controls their execution in order to get a good approximation of their continuous time behavior. Also a very simple model of an A/D converter has been introduced in order to model quantization, sampling, and delay of the converter. The converter still communicates via the interface of the signal. Note that there is no D/A converter — it is not mandatory for a simulation, but it can be added like the A/D converter.

11.4.3 Refinement of Module Interfaces

By the refinement of computation we have specified the points in time and the order in which the single blocks are computed, and to what precision they communicate. However, we have used an abstract interface for this communication that is unique for analog and digital blocks. In order to start the design in the separate partitions (circuit design or digital system design), we specify concrete, 'pin-accurate' interfaces. The basic idea of this refinement is already known from SystemC design methodology [OSCI, 2002b, Siegmund and Müller, 2001]: Adapter classes translate the abstract interface to a concrete, realistic pin and cycle accurate interface.

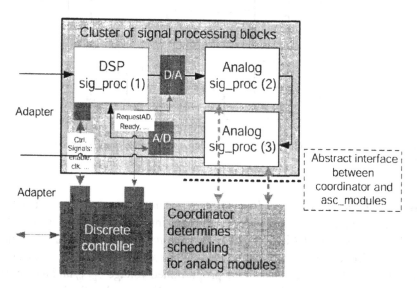

Figure 11.7. In this pin accurate model the abstract communication and synchronization is translated to 'physical' signals by adapters

Figure 11.7 shows the resulting structure. The modules have additional interfaces which are provided by adapter classes. These adapter classes translate the abstract communication to a concrete 'pin accurate' interface. Note that nevertheless both types of communication can be used. Compared to the refinement of communication/adapter classes known, these adapters provide a translation between a simulator coupling interface and signals, which explicitly control the discretization in a simulator.

11.5   A Simple Example for Modeling and Refinement

The SystemC classes presented and the refinement/evaluate methodology have been evaluated in a case study. In the case study the system shown in figure 11.8 has been designed.

The design example is a control loop which controls a continuous voltage $U(C)$ on a capacitor by switching two transistors T_1 or T_2 either on or off. When T_1 is on and T_2 is off the capacitor is charged via the resistor R. When T_1 is off and T_2 is on the capacitor is uncharged via the resistor R. The concept of the design is to use a control loop to bring $U(C)$ to a programmed voltage U_{prog} and to keep it there by switching transistors on or off with variable pulse width ('PWM controller').

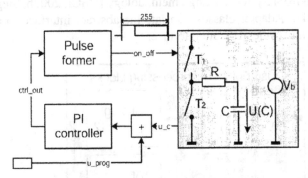

Figure 11.8. PWM controller example: The mathematical model used as executable specification

Mathematical Model.   In order to validate the system concept a first mathematical model can be created in a way as interactive and easy as in MAT-LAB/Simulink: The CMOS switches and the resistor/capacitor are modeled by a transfer function asc_spartial. A signal processing block asc_sub computes the difference between the programmed voltage U_{prog} and the actual voltage $U(C)$. The modules asc_s_pi_controller models the PI controller. Transfer function, adder, and PI controller are a signal processing cluster, which is controlled by a coordinator:

```
...
asc_coordinator      *c;
```

```
asc_spartial        *load;
asc_adder           *adder;
asc_s_pi_controller *controller;
asc_signal<double>  Uc, difference, ctrl_out, on_off;
...
```

The modules get the coordinator as a parameter when they are instantiated in the constructor of the module pwm:

```
SC_CTOR(pwm)
{
  ctrl_out.lb=0.0; ctrl_out.ub=255;
  c = new asc_coordinator(1,SC_US);

  load = new asc_spartial("load", c);
  load->add_pole(255, -1/0.05);
  load->x(on_off);
  load->y(Uc);

  adder = new asc_sub("adder", c);
  adder->a(Uc);
  adder->b(Uprog);
  adder->y(difference);

  controller = new asc_s_pi_controller("pi_controller", c);
  controller->k=0.001; controller->T=10;
  controller->x(difference); controller->y(ctrl_out);

  SC_THREAD(pulse_generator);
};
```

The coordinator activates all signal processing functions of his cluster in the step width specified as parameter (here: $1\mu s$); for the proof of concept we have chosen a very small step width, which is certainly good enough. Note that the class asc_signal used in the case study also synchronizes the signal processing cluster after a read and before a write from the discrete event domain. Section 11.3 only indicates this. Later we try to reduce the step width. The behavior of the pulse former is specified by a (discrete) SystemC thread (SC_THREAD):

```
void pulse_generator() {
  do {
    on_off = 1; wait(sc_time(ctrl_out.read(), SC_US));
    on_off = 0; wait(sc_time(255-ctrl_out.read(), SC_US));
  } while (true);
}
```

The complete listing of the example is in the appendix. Figure 11.9 shows a step response simulated with the executable specification. The executable

318

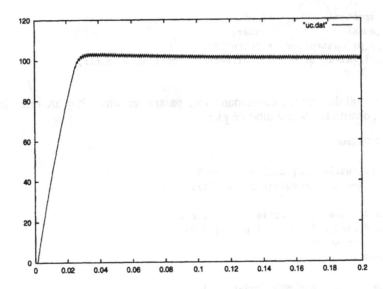

Figure 11.9. Step response of the PWM example

specification which specifies the ideal dynamical behavior nearly fits on just one page. Nevertheless, after full refinement one ends up with some hundred pages of source code for a complete controller. In the following, we give an overview of some of the refinement steps.

Some Refinements.   In the beginning of the design process we started with very small step widths which modeled the ideal dynamical behavior. Because a digital implementation of the controller is intended, we tried to enlarge the step width as much as possible. First, this is done easily by reducing the sample frequency of the controller. A further reduction is possible if the points for the sampling of the analog signals are specified more precisely. We modify this point in time by setting the sample time of the coordinator to infinitely. Then the coordinator only synchronizes after writes and before reads of the discrete event domain, if a synchronization is requested explicitly. This explicit synchronization then models the sampling of an A/D converter:

```
...
on_off = 1; wait(sc_time(ctrl_out.read()/2, SC_US));
coordinator->request_synchronization(0,0);
wait(sc_time(ctrl_out.read()/2, SC_US));
on_off = 0; wait(sc_time(255-ctrl_out.read(), SC_US));
...
```

The coordinator is now requested explicitly to perform a synchronization, which consists of computing actual values for the analog behavioral model of T_1, T_2, R, C, the adder, and the PI controller. This refinement step is very typical for a refine/evaluate flow: the refinement step requires only a very small modeling effort and is validated immediately.

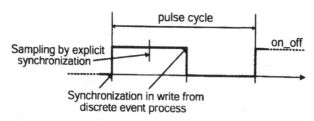

Figure 11.10. Refinement of the scheduling of operations in the control loop

After validating the functional correctness of this design step we replace the signal U_C by a rather complex model of an A/D converter. Furthermore, we replace the continuous time PI controller by a discrete time PI controller, which now works with a fixed sample rate. We also introduce a new block `controller` which explicitly calls the signal processing functions of the components to be realized in digital: the adder; and the PI controller. The component to be realized in analog remains under the control of a coordinator. We also validate the impact of quantization and limitation to the behavior of the system: The corresponding values can be modeled easily as properties of each signal. This also applies to signals and states in each block, especially the discrete controller.

Up to now we have most notably modeled the partitioning of functions to analog or digital blocks, quantization, and the timing/scheduling of operations. By the *refinement of interfaces* we add a number of ports to each block. Note that this refinement no longer modifies the overall system behavior. It just specifies an implementation of the functional description evaluated in the refinement of computation. In the given example given the discrete controller then realizes:

- A bus interface via which it can communicate with other components;

- A register map that stores a set of programmable parameters that can be accessed via the bus interface;

- A controller, that gets a high clock frequency and that generates clock- and enable signals for all digital blocks:

  - the A/D converter;

  - the adder;

  - the PI controller, and

  - the pulse former.

Figure 11.11 shows the detailed design of the PWM controller. Note that the simulation of the discrete controller and the detailed communication reduces simulation performance by magnitudes in exchange for modeling communication pin accurate.

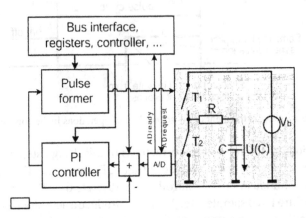

Figure 11.11. Refined PWM model

The design shown in figure 11.11 is the starting point for the synthesis of the digital modules at RT level, respectively, for the circuit level design of the analog part.

11.6   Conclusions

The methods described permit the modeling of analog and signal processing systems in SystemC 2.0 at a high level of abstraction, which can be compared to Simulink. Furthermore, they enable the design of such systems by a refine/evaluate approach. Features required for such a design methodology are:

- Signals which allow designers to connect non-conservative, continuous time signals with discrete time or discrete event signals in a direct way;

  This supports the interactive design, because there is no need to model converters explicitly. Furthermore, this allows the designer to model one part of the design by a rather abstract block diagram with non-conservative signals, connected with a design of other blocks which already use discrete event signals of SystemC 2.0;

- A well defined and generic interface between coordinator and signal processing blocks permits one to refine the computation. In this context the use of relaxation methods for distributed simulation permits designers to combine different (and changing!) models of computation for the modeling of a system on different levels of abstraction.

The refinement of mixed-signal systems is through going from a mathematical model down to behavior or register transfer level for the digital domain. It can be seen as an extension of hardware/software co-design to the domain of signal processing systems, where it also covers the design of analog/digital systems ('analog/digital co-design'). Most notably it closes the gap between the method oriented design of signal processing methods using Matlab/Simulink, for example and the hardware oriented design of mixed-signal systems using VHDL-AMS, for example.

A very demanding task which is not covered by this article is the refinement of analog modules with conservative signals (physical quantities). For the refinement of such modules we have to provide means that allow designers to couple non-conservative signals with conservative signals.

Acknowledgments

The author would like to thank Alain Vachoux, Karsten Einwich, Peter Schwarz, and Klaus Waldschmidt for many interesting and valuable discussions.

Appendix: Model of the PWM Controller

```
#include "systemc.h"
#include "asc.h"

//
// Case study:
// Design of a PWM controller by refinement.
// The following is the executable specification:
//
SC_MODULE(pwm)
{
  sc_in<double> Uprog;

  asc_coordinator      *c;
  asc_spartial         *load;
  asc_adder            *adder;
  asc_s_pi_controller  *controller;

  asc_signal<double>   Uc, difference, ctrl_out, on_off;

  // This part is modeled easily as a SC_THREAD:
  // It generates a discrete signal that changes its value at
  // two points in time that are computed.
  void pulse_generator()
  {
    do {
      sc_time on_time  = 5*sc_time(ctrl_out.read(), SC_US);
      sc_time off_time = 5*sc_time(255.0-ctrl_out.read(), SC_US);
      on_off.write(1); wait(on_time);
      on_off.write(0); wait(off_time);
    } while (true);
  }

  SC_CTOR(pwm)
  {
    ctrl_out.lb=0.0;
    ctrl_out.ub=255;
    c = new asc_coordinator(1,SC_MS);
    c->register_signal(&Uc);
    c->register_signal(&difference);
    c->register_signal(&ctrl_out);
    c->register_signal(&on_off);

    load = new asc_spartial("load", c);
    load->add_pole(255, -1/0.05);
    load->x(on_off);
    load->y(Uc);

    adder = new asc_sub("adder", c);
```

```
    controller = new asc_s_pi_controller("pi_controller", c);
    controller->k=0.001;
    controller->T=10;
    controller->x(difference);
    controller->y(ctrl_out);

    SC_THREAD(pulse_generator);
  }
};

int sc_main(int argc, char* argv[])
{
  // setup of simulator
  sc_set_time_resolution(1.0, SC_NS);
  sc_set_default_time_unit(1.0, SC_NS);

  sc_signal<double> Uprog;

  // From outside, nobody will ever see, that we have used
  // ASC components. Just instanciate it.
  pwm pwm("pwm1");
  pwm.Uprog(Uprog);
  Uprog = 100;
  sc_start(sc_time(0.5,SC_SEC));
  return 0;
}
```

References

[Abbaspour and Zhu, 2002] Abbaspour, M. and Zhu, J. (2002). Retargetable Binary Utilities. In *Proc. of the Design Automation Conference (DAC'02)*, New Orleans, USA. IEEE CS Press, Los Alamitos.

[Abraham and Agarwal, 1985] Abraham, J.A. and Agarwal, V.K. (1985). *Test Generation for Digital Systems — Fault-Tolerant Computing: Theory and Techniques*. Prentice-Hall, Englewood Cliffs.

[Abraham and Fuchs, 1986] Abraham, J.A. and Fuchs, K. (1986). Fault and Error Models for VLSI. In *Proc. of the IEEE*, 74(5).

[Accellera, 2002a] Accellera (2002a). Homepage — Accellera Interface Technical Committee. www.eda.org/itc.

[Accellera, 2002b] Accellera (2002b). *SytemVerilog 3.0 — Accellera's Extensions to Verilog(R)*. Accellera Organization, Inc.

[Allan et al., 2002] Allan, A., Edenfeld, D., Joyner, W., Kahng, A., Rodgers, M., and Zorian, Y. (2002). 2001 Technology Roadmap for Semiconductors. *IEEE Computer*, 35(1).

[ARC Int., 2001] ARC Int. (2001). The ARCtangent^{TM} A4 Core — A Technical Summary. www.arc.com/downloads/downloads-data-sheets.htm.

[ARM, 2002] ARM (2002). Homepage. www.arm.com.

[Armstrong et al., 1992] Armstrong, J.R., Lam, F.S., and Ward, P.C. (1992). Test Generation and Fault Simulation for Behavioral Models — Performance and Fault Modeling with VHDL. In Shoen, J.M., editor, *Performance and Fault Modelling with VHDL*. Prentice Hall, Englewood Cliffs.

[Ashenden et al., 1998] Ashenden, P., Wilsey, P., and Martin, D. (1998). SUAVE: Object-Oriented and Genericity Extensions to VHDL for High-Level Modeling. In *Proc. of the Forum on Specification & Design Languages (FDL'98)*, Lyon, France.

325

326

[Ashenden, 2002] Ashenden, P. J. (2002). *The Designer's Guide to VHDL*. Morgan Kaufmann Publishers, San Francisco.

[Ashenden et al., 2001] Ashenden, P. J., Biniasch, R., Fandrey, T., Grimpe, E., Schubert, A., and Schubert, Th. (2001). Input Language Subset Specification (formal). Technical Report, Kuratorium OFFIS e.V. IST FP5 Project ODETTE Deliverable R3.3/D3.1.

[Ashenden et al., 1997] Ashenden, P. J., Wilsey, P. A., and Martin, D. E. (1997). SUAVE Proposal for Extensions to VHDL for High-Level Modeling. Technical Report TR-97-07, The University of Adelaide, Adelaide, Australia.

[Balarin et al., 1997] Balarin, F., Giusto, P., Jurecska, A., Passerone, C., Sentovich, E., Tabbara, B., Chiodo, M., Hsieh, H., Lavagno, L., Sangiovanni-Vincentelli, A.L., and Suzuki, K. (1997). *Hardware-Software Co-Design of Embedded Systems*. Kluwer Academic Publishers, Boston/London/Dordrecht.

[Barnes, 1998] Barnes, J. (1998). *Programming in Ada95*. Addison-Wesley, Boston.

[Bhasker, 2002] Bhasker, J. (2002). *A SystemC Primer*. Star Galaxy Publishing, Allentown.

[Boehm, 1988] Boehm, B.W. (1988). A Spiral Model of Software Development and Enhancement. *IEEE Computer*, 21(5).

[Bonnerud et al., 2001] Bonnerud, T. E., Hernes, B., and Ytterdal, T. (2001). A Mixed-Signal, Functional Level Simulation Framework Based on SystemC System-On-a-Chip Applications. In *Proc. of the 2001 Custom Integrated Circuts Conference (CICC'01)*, San Diego, USA. IEEE CS Press, Los Alamitos.

[Booch, 1993] Booch, G. (1993). *Object-Oriented Analysis and Design With Applications*. Addison-Wesley, Boston.

[Börger et al., 1995] Börger, E., Glässer, U., and Müller, W. (1995). Formal Definition of an Abstract VHDL'93 Simulator by EA-Machines. In Delgado Kloos, C. and Breuer, P. T., editors, *Formal Semantics for VHDL*. Kluwer Academic Publishers, Boston/London/Dordrecht.

[Börger and Schulte, 1998] Börger, E. and Schulte, W. (1998). Defining the Java Virtual Machine as Platform for Provably Correct Java Compilation. In Brim, L., Gruska, J., and Zlatuska, J., editors, *Mathematical Foundations of Computer Science, MFCS 98*, Lecture Notes in Computer Science. Springer Verlag, Berlin/Heidelberg/New York.

[Börger and Stärk, 2003] Börger, E. and Stärk, R. (2003). *Abstract State Machines - A Method for High-Level System Design and Analysis*. Springer Verlag, Berlin/Heidelberg/New York.

[Burns and Wellings, 1998] Burns, A. and Wellings, A. (1998). *Concurrency in Ada*. Cambridge University Press, Cambridge.

[Cadence Design Systems, 2002] Cadence Design Systems (2002). Virtual Component Co-Design (VCC). www.cadence.com/datasheets.

[Carroll and Ellis, 1995] Carroll, M. D. and Ellis, M. A. (1995). *Designing and Coding Reusable C++*. Addison-Wesley, Boston.

[Cesario et al., 2002] Cesario, W., Baghdadi, A., Gauthier, L., Lyonnard, D., Nicolescu, G., Paviot, Y., Yoo, S., Jerraya, A., and Diaz-Nava, M. (2002). Component-Based Design Approach for Multicore SoCs. In *Proc. of the Design Automation Conference (DAC'02)*, New Orleans, USA. IEEE CS Press, Los Alamitos.

[Chakraborty and Gosh, 1988] Chakraborty, T.J. and Gosh, S. (1988). On Behavior Fault Modeling for Combinational Digital Designs. In *IEEE International Test Conference (ITC'88)*, Washington, USA. IEEE CS Press, Los Alamitos.

[Chang et al., 1999] Chang, H., Cooke, L., Hunt, M., Martin, G., McNelly, A., and Todd, L. (1999). *Surviving the SoC Revolution: A Guide to Platform-Based Design*. Kluwer Academic Publishers, Boston/Dordrecht/London.

[Cheng et al., 1999] Cheng, K.T., Huang, S.Y., and Dai, W.J. (1999). Fault Emulation: A New Methodology for Fault Grading. *IEEE Trans. on Computer Aided Design*, 18(10).

[Cheng and Krishnakumar, 1996] Cheng, K.T. and Krishnakumar, A.S. (1996). Automatic Generation of Functional Vectors Using the Extended Finite State Machine Model. *ACM Trans. on Design Automation of Electronic Systems*, 1(1).

[Chevallaz et al., 2002] Chevallaz, C., Mareau, N., and Gonier, A. (2002). Advanced Methods for SoC Concurrent Engineering. In *Proc. of Design, Automation and Test in Europe — Designer Forum (DATE'02)*, Paris, France. IEEE CS Press, Los Alamitos.

[Chiou et al., 1999] Chiou, D., Jain, P., Rudolph, L., and Devadas, S. (1999). Application-Specific Memory Management for Embedded Systems Using Software-Controlled Caches. In *Proc. of the Design Automation Conference (DAC'99)*, New Orleans, USA. IEEE CS Press, Los Alamitos.

[Civera et al., 2001] Civera, P., Macchiarulo, L., Rebaudengo, M., Reorda, M. Sonza, and Violante, M. (2001). Exploiting Circuit Emulation for Fast Hardness Evaluation. *IEEE Trans. on Nuclear Science*, 48(6).

[Clouard et al., 2002] Clouard, A., Mastrorocco, G., Carbognani, F., Perrin, A., and Ghenassia, F. (2002). Towards Bridging the Gap between SoC Transactional and Cycle-Accurate Levels. In *Proc. of Design, Automation and Test in Europe — Designer Forum (DATE'02)*, Paris, France. IEEE CS Press, Los Alamitos.

[Compaq et al., 2000] Compaq, Hewlett-Packard, Intel, Lucent, Microsoft, NEC, and Philips (2000). *Universal Serial Bus Specification, Revision 2.0.*

[Corno et al., 1997] Corno, F., Prinetto, P., and Reorda, M.S. (1997). Testability Analysis and ATPG on Behavioral RT-Level VHDL. In *IEEE International Test Conference (ITC'97)*, Washington, USA.

[CoWare, 2002] CoWare (2002). Homepage. www.coware.com.

[de Man, 2002] de Man, H. (2002). On Nanoscale Integration and Gigascale Complexity in the Post.com world. In *Design, Automation and Test in Europe (DATE'02)*, Paris, France. IEEE CS Press, Los Alamitos.

[Delgado Kloos and Breuer, 1995] Delgado Kloos, C. and Breuer, P. T. (1995). *Formal Semantics for VHDL.* Kluwer Academic Publishers, Boston/London/Dordrecht.

[Desmet et al., 2000] Desmet, D., Verkest, D., and de Man, H. (2000). Operating System Based Software Generation for Systems-On-Chip. In *Proc. of the Design Automation Conference (DAC'00)*, Los Angeles, USA. IEEE CS Press, Los Alamitos.

[Dey and Bommu, 1997] Dey, S. and Bommu, S. (1997). Performance Analysis of a System of Communicating Processes. In *Proc. of the International Conference on Computer Aided Design (ICCAD'97)*, San Jose, USA. IEEE CS Press, Los Alamitos.

[Einwich et al., 2001] Einwich, K., Clauss, Ch., Noessing, G., Schwarz, P., and Zojer, H. (2001). SystemC Extensions for Mixed-Signal System Design. In *Proc. of the Forum on Specification & Design Languages (FDL'01)*, Lyon, France.

[Einwich et al., 2002a] Einwich, K., Grimm, Ch., Vachoux, A., Martinez-Madrid, N., Moreno, F.R., and Ch., Meise (2002a). Analog and Mixed Signal Extensions for SystemC. White Paper of the OSCI SystemC-AMS Working Group.

[Einwich et al., 2002b] Einwich, K., Schwarz, P., Grimm, Ch., and Wald-schmidt, K. (2002b). Mixed-Signal Extensions for SystemC. In *Proc. of the Forum on Specification & Design Languages (FDL'02)*, Marseille, France.

[Ellervee et al., 2000] Ellervee, P., Miranda, M., Cathoor, F., and Hermani, A. (2000). System Level Data Format Exploration for Dynamically Allocated Data Structures. In *Proc. of the Design Automation Conference (DAC'00)*, Los Angeles, USA. IEEE CS Press, Los Alamitos.

[Ernst et al., 2002] Ernst, R., Richter, K., Haubelt, C., and Teich, J. (2002). System Design for Flexibility. In *Proc. of Design, Automation and Test in Europe (DATE'02)*, Paris, France. IEEE CS Press, Los Alamitos.

[ETSI, 2000a] ETSI (2000a). *Broadband Radio Access Networks (BRAN), HIPERLAN Type 2, Data Link Control (DLC) Layer, Part 1: Basic Data Transport Functions, ETSI TS 101 761-1.*

[ETSI, 2000b] ETSI (2000b). *Broadband Radio Access Networks (BRAN), HIPERLAN Type 2, System Overview, ETSI TR 101 683 (V1.1.2).*

[Fernández et al., 2002] Fernández, V., Herrera, F., Sánchez, P., and Villar, E. (2002). Conclusiones: Metodología Industrial Para Codiseño de Sistemas Embebidos HW/SW. Deliverable Document: DF FEDER 1FD97-0791, University of Cantabria, Cantabria, Spain.

[Ferrandi et al., 2002a] Ferrandi, F., Fummi, F., and Sciuto, D. (2002a). Test Generation and Testability Alternatives Exploration of Critical Algorithms for Embedded Applications. *IEEE Trans. on Computers*, 51(2).

[Ferrandi et al., 2002b] Ferrandi, F., Rendine, M., and Sciuto, D. (2002b). Functional Verification for SystemC Descriptions using Constraint Solving. In *Proc. of Design, Automation and Test in Europe (DATE'02)*, Paris, France. IEEE CS Press, Los Alamitos.

[Fin et al., 2002] Fin, A., Fummi, F., Galavotti, M., Pravadelli, G., Rossi, U., and Toto, F. (2002). Mixing ATPG and Property Checking for Testing HW/SW Interfaces. In *Proc. of the IEEE European Test Workshop (ETW'02)*, Corfu, Greece. IEEE CS Press, Los Alamitos.

[Fin et al., 2001a] Fin, A., Fummi, F., Martignano, M., and Signoretto, M. (2001a). SystemC: A Homogeneous Environment to Test Embedded Systems. In *Proc. of the IEEE International Symposium on Hardware/Software Codesign (CODES'01)*, Copenhagen, Denmark.

[Fin et al., 2001b] Fin, A., Fummi, F., and Pravadelli, G. (2001b). AMLETO: A Multi-Language Environment for Functional Test Generation. In *IEEE*

330

International Test Conference (ITC'01), Baltimore, USA. IEEE CS Press, Los Alamitos.

[Fornaciari et al., 2001] Fornaciari, W., Sciuto, D., Silvano, C., and Zaccaria, V. (2001). A Design Framework to Efficiently Explore Energy-Delay Tradeoffs. In *Proc. of the International Symposium on Hardware/Software Co-Design (CODES'01)*, Copenhagen, Denmark.

[Gajski et al., 1994a] Gajski, D., Vahid, F., and Narayan, S. (1994a). A System-Design Methodology: Executable-Specification Refinement. In *Proc. of the European Design Automation Conference (EURO-DAC'94)*, Paris, France. IEEE CS Press, Los Alamitos.

[Gajski et al., 1997] Gajski, D., Zhu, J., and Dömer, R. (1997). The SpecC+ Language. Technical Report ICS-TR-97-15, University of California, Irvine, Irvine, USA.

[Gajski et al., 2000] Gajski, D., Zhu, J., Dömer, R., Gerstlauer, A., and Zhao, S. (2000). *SpecC: Specification Language and Methodology*. Kluwer Academic Publishers, Boston/Dordrecht/London.

[Gajski et al., 1994b] Gajski, Daniel D., Vahid, Frank, and Narayan, Sanjiv (1994b). *Specification and Design of Embedded System*. Prentice Hall, Englewood Cliffs.

[Garnett, 2000] Garnett, N. (2000). EL/IX Base API Specification. DRAFT — V1.2. ELIX Specification.

[Gasteier and Glesner, 1999] Gasteier, M. and Glesner, M. (1999). Bus-Based Communication Synthesis on System Level. *ACM Transactions on Design Automation of Electronic Systems (TODAES)*, 4(1).

[Gauthier et al., 2001] Gauthier, L., Yoo, S., and Jerraya, A. (2001). Automatic Generation and Targeting of Application Specific Operating Systems and Embedded System Software. In *Proc. of Design, Automation and Test in Europe (DATE'01)*, Munich, Germany. IEEE CS Press, Los Alamitos.

[Gebotys, 2001] Gebotys, C. (2001). Utilizing Memory Bandwidth in DSP Embedded Processors. In *Proc. of the Design Automation Conference (DAC'01)*, Las Vegas, USA. IEEE CS Press, Los Alamitos.

[Gerlach and Rosenstiel, 2000] Gerlach, J. and Rosenstiel, W. (2000). System Level Design Using the SystemC Modeling Platform. In *Proc. of the Forum on Specification & Design Languages (FDL'00)*, Tübingen, Germany.

[Glässer et al., 1999] Glässer, U., Gotzhein, R., and Prinz, A. (1999). Towards a New Formal SDL Semantics Based on Abstract State Machines. *SDL '99 - The Next Millenium, 9th SDL Forum Proceedings*.

[Gosh, 1988] Gosh, S. (1988). Behavior-Level Fault Simulation. *IEEE Design &Test of Computers*, 5(3).

[Gosh and Chakraborty, 1991] Gosh, S. and Chakraborty, T.J. (1991). On Behavior Fault Modeling for Digital Designs. *Journal of Electronic Testing: Theory and Applications (JETTA)*, 2(2).

[Gracia et al., 2001] Gracia, J., Baraza, J.C., Gil, D., and Gil, P.J. (2001). Comparison and Application of Different VHDL-Based Fault Injection Techniques. In *Proc. of the IEEE International Symposium on Defect and Fault Tolerance in VLSI Systems (DFT'01)*, San Francisco, USA.

[Grimm et al., 2003] Grimm, Ch., Heupke, W., Meise, Ch., and Waldschmidt, K. (2003). Refinement of Mixed-Signal Systems with SystemC. In *Proc. of Design, Automation and Test in Europe (DATE'03)*, Munich, Germany. IEEE CS Press, Los Alamitos.

[Grimm et al., 2002] Grimm, Ch., Oehler, P., Meise, Ch., Waldschmidt, K., and Fey, W. (2002). AnalogSL: A Library for Modeling Analog Power Drivers with C++. In *System on Chip Design Languages*. Kluwer Academic Publishers, Boston/London/Dordrecht.

[Grimpe et al., 2002] Grimpe, E., Biniasch, R., Fandrey, T., Oppenheimer, F., and Timmermann, B. (2002). SystemC Object-Oriented Extensions and Synthesis Features. In *Proc. of the Forum on Specification & Design Languages (FDL'02)*, Marseille, France.

[Grötker, 2002] Grötker, T. (2002). Modeling Software with SystemC 3.0. In *6th European SystemC Users Group Meeting*.

[Grötker, 2002] Grötker, T. (2002). Reducing the SoC Design Cycle: Transaction-Level SoC Verification with SystemC. www.synopsys.com.

[Grötker et al., 2002] Grötker, T., Liao, S., Martin, G., and Swan, S. (2002). *System Design with SystemC*. Kluwer Academic Publishers, Boston/London/Dordrecht.

[Gupta, 1995] Gupta, R. (1995). *Co-Synthesis of Hardware and Software for Digital Embedded Systems*. Kluwer Academic Publishers, Boston/London/Dordrecht.

[Gupta et al., 2000] Gupta, S., Miranda, M., Catthoor, F., and Gupta, R. (2000). Analysis of High-level Address Code Transformations for Programmable

Processors. In *Proc. of Design, Automation and Test in Europe (DATE'00)*, Paris, France. IEEE CS Press, Los Alamitos.

[Gurevich, 1991] Gurevich, Y. (1991). Evolving Algebras. A Tutorial Introduction. *Bulletin of EATCS*, 43.

[Gurevich, 1995] Gurevich, Y. (1995). Evolving Algebras 1993: Lipari Guide. In Börger, E., editor, *Specification and Validation Methods*. Oxford University Press, Oxford.

[Harris and Zhang, 2001] Harris, I.G. and Zhang, Q. (2001). A Validation Fault Model for Timing-Induced Functional Errors. In *IEEE International Test Conference (ITC'01)*, Baltimore, USA.

[Hashmi and Bruce, 1995] Hashmi, M.M.K. and Bruce, A.C. (1995). Design and Use of a System Level Specification and Verification Methodology. In *Proc. of the European Design Automation Conference (EDAC'95)*, Brighton, UK.

[Haverinen et al., 2002] Haverinen, A., Leclercq, M., Weyrich, N., and Wingard, D. (2002). SystemC-Based SoC Communication Modeling for the OCP Protocol. www.ocpip.org.

[Herrera et al., 2003] Herrera, F., Posadas, H., Sánchez, P., and Villar, E. (2003). Systematic Embedded Software Generation from SystemC. In *Proc. of Design, Automation and Test in Europe (Date'03)*, Munich, Germany. IEEE CS Press, Los Alamitos.

[Hines and Borriello, 1997] Hines, K. and Borriello, G. (1997). Optimizing Communication in Embedded System Co-simulation. In *Proc. of the Fifth International Workshop on HW/SW Codesign*, Braunschweig, Germany.

[Husak, 2002] Husak, D. (2002). Network Processors: A Definition and Comparison. www.motorola.com/collateral/M957198397651.pdf.

[IEEE, 1999a] IEEE (1999a). *IEEE Std 1076.1-1999, VHDL-AMS*. IEEE Press, New York.

[IEEE, 1999b] IEEE (1999b). *Wireless LAN Medium Access Control (MAC) and Physical Layer (PHY) Specifications, ANSI/IEEE Std 802.11 1999 Edition*.

[Jacobson et al., 1992] Jacobson, I., Christerson, M., Jonsson, P., and Oevergaard, G. (1992). *Object-Oriented Software Engineering: A Use Case Driven Approach*. Addison-Wesley, Boston.

[Jain et al., 2001] Jain, P., Devadas, S., Engels, D., and Rupoldh, L. (2001). Software-Assisted Cache Replacement Mechanisms for Embedded System. In *Proc. of the International Conference on Computer Aided Design (IC-CAD'01)*, San Jose, USA. IEEE CS Press, Los Alamitos.

[Jenn et al., 1994] Jenn, E., Arlat, J., Rimen, M., Ohlsson, J., and Karlsson, J. (1994). Fault Injection into VHDL Models: The MEFISTO Tool. In *IEEE International Symposium on Fault-Tolerant Computing (FTCS'94)*, Seattle, USA.

[Jeruchim et al., 1992] Jeruchim, M.C., Balaban, P., and Shanmugan, K.S. (1992). *Simulation of Communication Systems*. Plenum Press, New York.

[Jiang and Brayton, 2002] Jiang, Y. and Brayton, R. (2002). Software Synthesis from Synchronous Specifications Using Logic Simulation Techniques. In *Proc. of the Design Automation Conference (DAC'02)*, New Orleans, USA. IEEE CS Press, Los Alamitos.

[K. Shumate, 1992] K. Shumate, M. Keller (1992). *Software Specification and Design*. John Wiley&Sons, Boston/Dordrecht/London.

[Keutzer et al., 2000] Keutzer, K., Malik, S., Newton, R., Rabacy, J., and Sangiovanni-Vincentelli, A. (2000). System Level Design: Orthogonalization of Concerns and Platform Based Design. *IEEE Trans. On Computer-Aided Design of Circuits and Systems*, 19(12).

[Kjeldsberd et al., 2001] Kjeldsberd, P. G., Cathoor, F., and Aas, E. (2001). Detection of Partially Simultaneously Alive Signal in Storage Requirement Estimation for Data Intensive Applications. In *Proc. of the Design Automation Conference (DAC'01)*, Las Vegas, USA. IEEE CS Press, Los Alamitos.

[Knudsen and Madsen, 1998] Knudsen, P.V. and Madsen, J. (1998). Integrating Communication Protocol Selection with Hardware/Software Codesign. In *Proc. of the International Symposium on System Level Synthesis (ISSS'98)*, Hsinchu, Taiwan.

[Knudsen and Madsen, 1999] Knudsen, P.V. and Madsen, J. (1999). Graph-Based Communication Analysis for Hardware/Software Codesign. In *Proc. of the International Workshop on HW/SW Codesign*, Rome, Italy.

[Kuhn et al., 1998] Kuhn, T., Rosenstiel, W., and Kebschull, U. (1998). Object Oriented Hardware Modeling and Simulation Based on Java. In *Proc. of the International Workshop on IP Based Synthesis and System Design*, Grenoble, France.

334

[Lahiri et al., 1999] Lahiri, K., Raghunathan, A., and Dey, S. (1999). Fast Performance Analysis of Bus-Based System-On-Chip Communication Architectures. In *Proc. of the International Conference on Computer Aided Design (ICCAD'99)*, San Jose, USA.

[Lahiri et al., 2001] Lahiri, K., Raghunathan, A., and Dey, S. (2001). System-Level Performance Analysis for Designing On-Chip Communication Architectures. *IEEE Transactions on Computer-Aided Design of Integrated Circuits and Systems*, 20(6).

[Lee, 2000] Lee, E.A. (2000). What's Ahead for Embedded Software? *IEEE Computer*, 33(9).

[Lee and Sangiovanni-Vincentelli, 1998] Lee, E.A. and Sangiovanni-Vincentelli, A. (1998). A Framework for Comparing Models of Computation. *IEEE Trans. On Computer-Aided Design of Integrated Circuits and Systems*, 17(12).

[Lekatsas et al., 2000] Lekatsas, H., Henkel, J, and Wolf, W. (2000). Code Compression as a Variable in Hardware/Software Co-Design. In *Proceeding of the International Symposium on Hardware/Software Co-Design (CODES'00)*, San Diego, USA.

[Leveugle, 2000] Leveugle, R. (2000). Fault Injection in VHDL Descriptions and Emulation. In *Proc. of the IEEE International Symposium on Defect and Fault Tolerance in VLSI Systems (DFT'00)*, Antwerpen, Belgium.

[Lieverse et al., 2001] Lieverse, P., Wolf, P.V., and Deprettere, E. (2001). A Trace Transformation Technique for Communication Refinement. In *International Workshop on HW/SW Codesign*, Copenhagen, Denmark.

[Liu et al., 1999] Liu, J., Wu, B., Liu, X., and Lee, E. A. (1999). Interoperation of Heterogeneous CAD Tools in Ptolemy II. In *Symposium on Design, Test and Microfabrication of MEMS/MOEMS (DTM'99)*, Paris, France.

[Lopez et al., 1993] Lopez, J.C., Hermida, R., and Geisselhardt, W. (1993). *Advanced Techniques for Embedded System Design and Test*. Kluwer Academic Publishers, Boston/London/Dordrecht.

[Lyons, 1997] Lyons, Richard G. (1997). *Understanding Digital Signal Processing*. Addison-Wesley, Boston.

[Malaiya et al., 1995] Malaiya, Y.K., Li, M.N., Bieman, J.M., and Karcich, R. (1995). Generation of Design Verification Tests from Behavioral VHDL Programs Using Path Enumeration and Constraint Programming. *IEEE Trans. on VLSI Systems*, 3(2).

[Marwedel and Goossens, 1995] Marwedel, P. and Goossens, G. (1995). *Code Generation for Embedded Processor*. Kluwer Academic Publishers, Boston/London/Dordrecht.

[Massa, 2002] Massa, A. (2002). *Embedded Software Development with eCos*. Prentice Hall, Englewood Cliffs.

[MathWorks, 1993] The MathWorks (1993). *The Matlab EXPO: An Introduction to MATLAB, SIMULINK, and the MATLAB Application Toolboxes*.

[MathWorks, 2000a] The MathWorks (2000a). *External Interfaces v6*.

[MathWorks, 2000b] The MathWorks (2000b). *Using Matlab v6*.

[McMillan, 1993] McMillan, K.L. (1993). *Symbolic Model Checking*. Kluwer Academic Publishers, Boston/London/Dordrecht.

[Mooney and Blough, 2002] Mooney, V. and Blough, D. (2002). A Hardware-Software Real-Time Operating System Framework for SoCs. *IEEE Design and Test of Computers*, 19(6).

[Moskewicz et al., 2001] Moskewicz, M., Madigan, C., Zhao, Y., Zhang, L., and Malik, S. (2001). Chaff: Engineering an Efficient SAT Solver. In *Design Automation Conference (DAC'01)*, Las Vegas, USA. IEEE CS Press, Los Alamitos.

[Müller et al., 2002] Müller, W., Dömer, R., and Gerstlauer, A. (2002). The Formal Execution Semantics of SpecC. In *Proc. of the International Symposium on System Synthesis (ISSS'02), Kyoto, Japan*.

[Müller et al., 2001] Müller, W., Ruf, J., Hofmann, D., Gerlach, J., Kropf, Th., and Rosenstiel, W. (2001). The Simulation Semantics of SystemC. In *Proc. of Design, Automation and Test in Europe (DATE'01)*, Munich, Germany. IEEE CS Press, Los Alamitos.

[Myers, 1999] Myers, G.J. (1999). *The Art of Software Testing*. Wiley & Sons, New York.

[Nagel et al., 2001] Nagel, P., Leyh, M., and Speitel, M. (2001). Using Behavioral Compiler and FPGA Prototyping for the Development of an OFDM Demodulator. In *User Papers of Eleventh Annual Synopsys Users Group Conference (SNUG)*, San Jose, USA.

[Niemann, 2001] Niemann, B. (2001). SystemC 1.x Introduction & Tutorial. Forth European SystemC Users Group Meeting. www-ti.informatik.uni-tuebingen.de/systemc.

[Niemann and Speitel, 2001] Niemann, B. and Speitel, M. (2001). The Application of SystemC Compiler in an Industrial Project — Modeling a GPS Receiver Using SystemC. In *User Papers of Synopsys Users Group Europe (SNUG)*, Munich, Germany.

[Nikkei, 2001] Nikkei (2001). Performance Analysis of a SoC Interconnect using SysProbe. Nikkei Electronics Magazine.

[Öberg et al., 1996] Öberg, J., Kumar, A., and Hemani, A. (1996). Grammar-based hardware synthesis of data communication protocols. In *Proc. of the 9th International Symposium on System Synthesis (ISSS'96)*, La Jolla, USA.

[OMG, 2001] OMG (2001). *Unified Modeling Language Specification v1.4.* www.omg.org.

[OSCI, 2000] OSCI (2000). *SystemC Version 1.0 User's Guide.* Synopsys Inc, CoWare Inc, Frontier Inc. www.systemc.org.

[OSCI, 2002a] OSCI (2002a). *SystemC Version 2.0 Functional Specification.* Synopsys Inc, CoWare Inc, Frontier Inc. www.systemc.org.

[OSCI, 2002b] OSCI (2002b). *SystemC Version 2.0 User's Guide.* Synopsys Inc, CoWare Inc, Frontier Inc. www.systemc.org.

[OSCI, 2002a] OSCI (2002a). *SystemC Version 2.0 User's Guide. Update for SystemC 2.0.1.* Open SystemC Initiative. www.systemc.org.

[OSCI, 2002b] OSCI (2002b). *Version 2.0.1 Master-Slave Communication Library - A SystemC Standard Library.* Open SystemC Initiative. www.systemc.org.

[OSCI, 2002c] OSCI WG Verification (2002c). SystemC Verification Standard Specification V1.0b. www.systemc.org.

[Passerone et al., 2001] Passerone, C., Watanabe, Y., and Lavagno, L. (2001). Generation of Minimal Size Code for Schedule Graphs. In *Proc. of Design, Automation and Test in Europe (DATE'01)*, Munich, Germany. IEEE CS Press, Los Alamitos.

[Pazos et al., 2002] Pazos, N., Brunnbauer, W., Foag, J., and Wild, T. (2002). System-Based Performance Estimation of Multi-Processing, Multi-Threading SoC Networking Architecture. In *Proc. of the Forum on Specification & Design Languages (FDL'02)*, Marseille, France.

[Petzold, 1983] Petzold, L. R. (1983). A Description of DASSL: A Differential-Algebraic System Solver. In Stepleman, R. S., editor, *Scientific Computing*. North-Holland, Amsterdam.

[Pop et al., 2000] Pop, P., Eles, P., and Peng, Z. (2000). Performance Estimation for Embedded Systems with Data and Control Dependencies. In *International Workshop on HW/SW Codesign*, San Diego, USA.

[Pross, 2002] Pross, U. (2002). Modellierung und Systemsimulation des Universal Serial Bus Protokoll. Master's thesis, Technische Universität Chemnitz, Chemnitz, Germany.

[Radetzki, 2000] Radetzki, M. (2000). *Synthesis of Digital Circuits from Object-Oriented Specifications*. PhD thesis, University of Oldenburg, Oldenburg, Germany.

[Radetzki et al., 1998] Radetzki, M., Putzke-Röming, W., and Nebel, W. (1998). Objective VHDL: The Object-Oriented Approach to Hardware Reuse. In Roger, J.-Y., Stanford-Smith, B., and Kidd, P. T., editors, *Advances in Information Technologies: The Business Challenge*, Amsterdam, Netherlands. IOS Press.

[Rational, 2001] Rational (2001). *Rational Unified Process Whitepaper: Best Practices for Software Development Teams*.

[Rowson and Sangiovanni-Vincentelli, 1997] Rowson, J.A. and Sangiovanni-Vincentelli, A. (1997). Interface-Based Design. In *Proc. of the Design Automation Conference (DAC'97)*, Anaheim, USA. IEEE CS Press, Los Alamitos.

[Salem and Shams, 2003] Salem, A. and Shams, A. (2003). The Formal Semantics of Synchronous SystemC. In *Proc. of Design, Automation and Test in Europe (DATE'03)*, Munich, Germany. IEEE CS Press, Los Alamitos.

[Sangiovanni-Vincentelli, 2002a] Sangiovanni-Vincentelli, A. (2002a). Defining Platform-Based Design. *EEDesign of EETimes*.

[Sangiovanni-Vincentelli, 2002b] Sangiovanni-Vincentelli, A. (2002b). The Context for Platform-Based Design. *IEEE Design & Test of Computer*, 19(6).

[Sasaki, 1999] Sasaki, H. (1999). A Formal Semantics for Verilog-VHDL Simulation Interoperability by Abstract State Machine. In *Proc. of Design, Automation and Test in Europe (DATE'99)*, Munich, Germany. IEEE CS Press, Los Alamitos.

[Sasaki et al., 1997] Sasaki, H., Mizushima, K., and Sasaki, T. (1997). Semantic Validation of VHDL-AMS by an Abstract State Machine. In *IEEE/VIUF International Workshop on Behavioral Modeling and Simulation (BMAS'97)*, Arlington, USA.

338

[Sayinta et al., 2003] Sayinta, A., Canverdi, G., Pauwels, M., Alshawa, A., and Dehaene, W. (2003). A Mixed Abstraction Level Co-Simulation Case Study Using SystemC for System-On-Chip Verification. In *Proc. of Design, Automation and Test in Europe (DATE'03)*, Munich, Germany. IEEE CS Press, Los Alamitos.

[Scanlon, 2002] Scanlon, T. (2002). Global Responsibilities in System-On-Chip Design. In *Proc. of Design, Automation and Test in Europe (DATE'02)*, Paris, France. IEEE CS Press, Los Alamitos.

[Schumacher, 1999] Schumacher, G. (1999). *Object-Oriented Hardware Specification and Design with a Language Extension to VHDL*. PhD thesis, University of Oldenburg, Oldenburg, Germany.

[Semeria and Ghosh, 1999] Semeria, L. and Ghosh, A. (1999). Methodology for Hardware/Software Co-verification in C/C++. In *Proc. of the High-Level Design Validation and Test Workshop (HLDVT'99)*, Oakland, USA.

[Sgroi et al., 1999] Sgroi, M., Lavagno, L., Watanabe, Y., and Sangiovanni-Vincentelli, A. (1999). Synthesis of Embedded Software Using Free-Choise Petri Nets. In *Proc. of the Design Automation Conference (DAC'99)*, New Orleans, USA. IEEE CS Press, Los Alamitos.

[Shiue, 2001] Shiue, W.T. (2001). Retargeable Compilation for Low Power. In *Proc. of the International Symposium on Hardware/Software Co-Design (CODES'01)*, Copenhagen, Denmark.

[Siegmund and Müller, 2001] Siegmund, R. and Müller, D. (2001). SystemC-SV: Extension of SystemC for Mixed Multi Level Communication Modeling and Interface-based System Design. In *Proc. of Design, Automation and Test in Europe (DATE'01)*, Munich, Germany. IEEE CS Press, Los Alamitos.

[Siegmund and Müller, 2002] Siegmund, R. and Müller, D. (2002). A Novel Synthesis Technique for Communication Controller Hardware from declarative Data Communication Protocol Specifications. In *Proc. of the Design Automation Conference (DAC'02)*, New Orleans, USA. IEEE CS Press, Los Alamitos.

[Sieh et al., 1997] Sieh, V., Tschäche, O., and Balbach, F. (1997). VERIFY: Evaluation of Reliability Using VHDL-Models with Embedded Fault Descriptions. In *IEEE International Symposium on Fault-Tolerant Computing (FTCS'97)*, Seattle, USA.

[Sigwarth et al., 2002] Sigwarth, C., Mayer, F., Schlicht, M., Kilian, G., and Heuberger, A. (2002). ASIC Implementation of a Receiver Chipset for the Broadcasting in Long-, Medium- and Shortwave Bands (DRM). In *Proc.*

of the International Symposium on Consumer Electronics (ISCE'02), Erfurt, Germany.

[Sirius Satellite Radio, 2003] Sirius Satellite Radio (2003). Homepage. www. cdradio.com.

[Soininen et al., 1995] Soininen, J.P., Huttunen, T., Tiensyrja, K., and Heusala, H. (1995). Cosimulation of Real-Time Control Systems. In *Proc. of the European Design Automation Conference (EURO-DAC'95)*, Brighton, UK.

[Straunstrup and Wolf, 1997] Straunstrup, J. and Wolf, W. (1997). *Hardware/-Software Co-Design: Principles and Practice*. Kluwer Academic Publishers, Boston/London/Dordrecht.

[Sung and Kum, 1995] Sung, W. and Kum, K. (1995). Simulation-Based Word-Length Optimization Method for Fixed-Point Digital Signal Processing Systems. *IEEE Transactions on Signal Processing*, 3(12).

[Suzuki and Sangiovanni-Vincentelli, 1996] Suzuki, K. and Sangiovanni-Vincentelli, A. (1996). Efficient Software Performance Estimation Methods for Hardware/Software Codesign. In *Proc. of the Design Automation Conference (DAC'96)*, Las Vegas, USA. IEEE CS Press, Los Alamitos.

[Swamy et al., 1995] Swamy, S., Molin, A., and Covnot, B. (1995). OO-VHDL Object-Oriented Extensions to VHDL. *IEEE Computer*, 28(10).

[Synopsys, 2002a] Synopsys Inc. (2002a). *CoCentric(R) SystemC Compiler Behavioral Modeling Guide*. www.synopsys.com.

[Synopsys, 2002b] Synopsys Inc. (2002b). *CoCentric(R) SystemC Compiler RTL User and Modeling Guide*. Synopsys, Inc. www.synopsys.com.

[Synopsys, 2002c] Synopsys Inc. (2002c). *Guide to Describing Synthesizable RTL in SystemC*. www.synopsys.com/sld.

[Synopsys, 1998] Synopsys Inc. (1998). *Protocol Compiler User Manual*. www.synopsys.com.

[Thiele et al., 2002] Thiele, L., Chakraborty, S., Gries, M., and Künzli, S. (2002). A Framework for Evaluating Design Tradeoffs in Packet Processing Architectures. In *Proc. of the Design Automation Conference (DAC'02)*, New Orleans, USA. IEEE CS Press, Los Alamitos.

[Thomas and Moorby, 1991] Thomas, D. E. and Moorby, P. R. (1991). *The Verilog Hardware Description Language*. Kluwer Academic Publishers, Boston/London/Dordrecht.

340

[Turner, 1993] Turner, K. J., editor (1993). *Using Formal Description Techniques. An Introduction to ESTELLE, LOTOS and SDL.* John Wiley & Sons, New York.

[University of Cincinnati, 1999] University of Cincinnati, Dept. of ECECS (1999). Savant Programmer's Manual.

[Vachoux et al., 2003a] Vachoux, A., Grimm, Ch., and Einwich, K. (2003a). Analog and Mixed-Signal Modeling with SystemC-AMS. In *International Symposion on Circuits and Systems 2003 (ISCAS '03)*, Bangkok, Thailand.

[Vachoux et al., 2003b] Vachoux, A., Grimm, Ch., and Einwich, K. (2003b). SystemC-AMS Requirements, Design Objectives and Rationale. In *Proc. of Design, Automation and Test in Europe (DATE'03)*, Munich, Germany. IEEE CS Press, Los Alamitos.

[Valderrama et al., 1996] Valderrama, C.A., Nacabal, F., Paulin, P., Jerraya, A.A., Attia, M., and Cayrol, O. (1996). Automatic Generation of Interfaces for Distributed C-VHDL Cosimulation of Embedded Systems: An Industrial Experience. In *IEEE Workshop on Rapid System Prototyping (RSP'96)*, Thessaloniki, Greece.

[van Nee and Prasad, 2000] van Nee, R. and Prasad, R. (2000). *OFDM for Wireless Multimedia Communications.* Artech House Publishers, Boston.

[Vanderperren et al., 2002] Vanderperren, Y., Dehaene, W., and Pauwels, M. (2002). A Method for the Development of Combined Floating- and Fixed-Point SystemC Models. In *Proc. of the Forum on Specification & Description Languages (FDL'02)*, Marseille, France.

[Vanmeerbeeck et al., 2001] Vanmeerbeeck, G., Schaumont, P., Vernalde, S., Engels, M., and Bolsens, I. (2001). Hardware/Software Partitioning for Embedded Systems in OCAPI-xl. In *Proc. of the IEEE International Symposium on Hardware/Software Codesign (CODES'01)*, Copenhagen, Denmark.

[Vercauteren et al., 1996] Vercauteren, S., Lin, B., and de Man, H (1996). A Strategy for Real-Time Kernel Support in Application-Specific HW/SW Embedded Architectures. In *Proc. of the Design Automation Conference (DAC'96)*, Las Vegas, USA. IEEE CS Press, Los Alamitos.

[Vernalde et al., 1999] Vernalde, S., Schaumont, P., and Bolsens, I. (1999). An Object Oriented Programming Approach for Hardware Design. In *IEEE Workshop on VLSI*, Orlando, USA.

[Villar, 2001] Villar, E. (2001). *Design of Hardware/Software Embedded Systems.* Servicio de Publicaciones de la Universidad de Cantabria, Cantabria, Spain.

[Villar, 2002] Villar, E. (2002). A Framework for Specification and Verification of Timing Constraints. In *System on Chip Design Languages*. Kluwer Academic Publishers, Boston/London/Dordrecht.

[Villar and Sánchez, 1993] Villar, E. and Sánchez, P. (1993). Synthesis Applications of VHDL. In Mermet, J.P., editor, *Fundamentals and Standards in Hardware Description Languages*. Kluwer Academic Publishers, Boston/London/Dordrecht.

[Wallace, 1995] Wallace, C. (1995). The Semantics of the C++ Programming Language. In Börger, E., editor, *Specification and Validation Methods*. Oxford University Press, Oxford.

[Ward and Mellor, 1985] Ward, P. and Mellor, S. (1985). *Structured Development for Real-Time Systems*. Yourdon Press, Englewood Cliffs.

[WindRiver, 2002] WindRiver (2002). Homepage. www. wrs.com.

[XM Satellite Radio, 2003] XM Satellite Radio (2003). Homepage. www. xmsatradio.com.

[Yu et al., 2002] Yu, X., A. Fin, F. Fummi, and Rudnick, E.M. (2002). A Genetic Testing Framework for Digital Integrated Circuits. In *IEEE International Conference on Tools for AI (ICTAI'02)*, Washington, USA.

[Zojer et al., 2000] Zojer, B., Koban, R., Pichler, J., and Paoli, G. (2000). A Broadband High-Voltage SLIC for a Splitter- and Transformerless Combined ADSL-Lite/POTS Linecard. *IEEE Journal of Solid-State Circuits*, 35(12).

Index

A4, 65
    See also ARCtangent-A4
Abstract State Machine (ASM), 97, 99–101
    See also ASM
abstraction level, 31
    algorithmic model, 34
    behavioral, 218
    cycle accurate (CA), 8, 10–11, 21, 58
    mixing, 8, 10
    refinement, 8, 10
    register transfer, 218, 258, 270
    system, 251–252, 260
    timed functional (TF), 8, 10–11, 21
    transaction level model (TLM), 33
    untimed functional (UTF), 7
AC-domain, 294
AHB, 59
algebra, 98
algorithmic model, 34
AMLETO, 134
analog
    and mixed-signal, 274
    filter, 276
    solver, 292
API
    operating system, 252
    RTOS, 261, 265–266, 270, 272
architecture, 31
ARCtangent-A4, 80, 83–84, 87
    ALU, 89, 93
    data path, 89–90, 93
    extensions, 87–88
ARM, 59
array mapping, 236
ASC library, 301, 305
ASCSG, 159, 162
    access node, 169
    accuracy, 189
    architectural model, 163, 175
    communication model, 164, 176
    functional model, 163, 174
    implementation, 173
    re-building effort, 189
    shared element, 169

    speed, 189
    verification, 184
ASM, 97, 99–101
    See also Abstract State Machine
    agent, 101
    CHOOSE, 101
    EXTEND, 100
    guarded function update, 99
    RANGESOVER, 101
    rules, 99
    VAR, 101
ATPG, 128, 136, 140, 148
automation, 179
    configuration file, 173, 182
    data structure, 179
automotive application, 275

back annotation, 42
back tracking, 278
backward Euler, 283
base band simulation, 290
behavioral level, 218
bilinear transformation, 283
bit coverage, 130
block diagram, 276, 301
BPSK, 67
branch coverage, 133
buffer
    input, 80–81
    output, 80–81

cancel, 108
capacitor, 296
carrier frequency, 290
channel, 51, 104, 192, 211, 255, 260, 267
    abstraction levels, 8
    design, 8
    dynamic creation, 252
    emulation, 232
    primitive, 8, 106–107
    sc_fifo, 222
    standard primitive, 256
    SW implementation, 267
    synthesis, 222

346